Fault Detection and Isolation

Nader Meskin • Khashayar Khorasani

Fault Detection and Isolation

Multi-Vehicle Unmanned Systems

Nader Meskin
Department of Electrical Engineering
College of Engineering
Qatar University
PO Box 2713, Doha, Qatar
nader.meskin@qu.edu.qa

Khashayar Khorasani
Concordia University
Dept. Electrical & Computer
Engineering
1455 de Maisonneuve Blvd. West
H3G 1M8 Montreal Québec
Canada
kash@ece.concordia.ca

ISBN 978-1-4899-8215-5   ISBN 978-1-4419-8393-0 (eBook)
DOI 10.1007/978-1-4419-8393-0
Springer New York Dordrecht Heidelberg London

Softcover re-print of the Hardcover 1st edition 2011

Printed on acid-free paper

Springer is part of Springer Science+Business Media (www.springer.com)

To Mojdeh, my wife
and
to Moslem and Soraya, my parents
for their unwavering love
Nader Meskin

To My Family

Khashayar Khorasani

Preface

Recent technological advances have generated a broad interest in cooperative control, management, and monitoring of systems composed of autonomous unmanned vehicles such as spacecraft, unmanned aerial vehicles, autonomous underwater vehicles, automated highway systems and multiple mobile robots. In these systems, a vehicle can interact autonomously with environment and other vehicles to perform tasks that are beyond the capabilities of individual vehicles. The ultimate ambitious goal of these systems is to achieve *autonomy*, representing the capability of these networked vehicles to accomplish the mission goals in face of significant uncertainty and unexpected events such as faults all with minimal or without human intervention. Unmanned vehicles such as unmanned aerial vehicles (UAVs) are indeed among the most complicated systems that are being developed with applications to both civilian and military domains. Since these vehicles operate in an environment subjected to a high degree of uncertainties and disturbances, the problem of precise and accurate control and estimation of these vehicles is difficult and requires advanced control and estimation theories. On the other hand, with an increasing requirement for control systems to be more secure and reliable, fault tolerance in such control systems is becoming more and more critical and significant.

In presence of undesirable effects such as failures in the actuators or sensors, the vehicles control systems must be responsive and adaptive to such failures. Under these circumstances there may be a necessity to adjust the control laws to recover the vehicles from the effects of anomalies and failures. Furthermore, the required adjustments and reconfiguration of control laws must be accomplished expeditiously in a relatively short period of time to guarantee the proper operation of the safety critical subsystems of the vehicles. Specifically, one is required to develop an autonomous fault diagnosis, health monitoring, and reconfigurable control systems. In particular for a network of unmanned vehicles, if any of the vehicles undergoes a failure or a sensor or actuator fault, and moreover the vehicle controller is not equipped with autonomous fault tolerant capabilities, the stability of the entire network may no longer be maintained which could lead to instability. Therefore, the problem of fault tolerant control systems for a network of unmanned vehicles needs to be further studied and investigated.

Advanced computers have resulted in more capable and sophisticated vehicle control systems. Consequently, activities such as guidance, navigation, and control; maneuver planning, command planning and sequencing; and fault diagnosis and recovery can all potentially be autonomously handled onboard the vehicle. Aside from the obvious cost savings that are realized by smaller operational staff, there are conventional advantages to placing some of these functions onboard the vehicle. For example, onboard fault diagnosis and recovery algorithms can detect, identify, and remedy vehicle faults, both minor and major, *in real-time*, possibly saving the mission in the process. Moreover, the ability to plan the vehicle activities onboard allows one to respond to major instrument failures or other anomalies without impacting the remaining healthy subsystems. In the event of a major anomaly, the vehicle can respond quickly and generate a new sequence of commands to carry out the remaining possible mission objectives.

Cooperative control and estimation of networked unmanned vehicles poses many interesting challenges such as optimal control of these vehicles, collision and obstacle avoidance during reconfiguration, information flow among the vehicles and development of advanced hardware devices such as sensors and actuators. One of the main challenges in these systems is developing autonomous cooperative control and estimation strategies that can maintain the group behavior and mission performance in presence of undesirable events such as failures in vehicles sensors, actuators, or other components. In order to achieve this goal, the team should have the capability of detecting and isolating faults in its vehicles and reconfigure the cooperative control algorithms to compensate for the faults. This book explores the Fault Detection and Isolation (FDI) problem in these systems.

This book is organized as follows. In Chapter 1, an overview of the fault detection and isolation literature, in general, and fault diagnosis for a network of unmanned vehicles, in particular, are provided. In Chapter 2, a brief review of Structured Fault Detection and Isolation Problem (SFDIP) for both linear and nonlinear systems are presented. A geometric FDI framework for solving the SFDIP problem is presented for linear and nonlinear systems in Sections 2.2 and 2.3, respectively.

In Chapter 3, we focus on the problem of fault detection and isolation in a network of unmanned vehicles without considering the effects of communication channels. It is shown in Section 3.1 that the actuator fault signatures in a network of unmanned vehicles with relative state measurements are dependent and the entire network can be considered as an overactuated system. This motivate us to develop a novel coding scheme applicable to systems with dependent fault signatures. Moreover, it is shown in Section 3.1.2 that the FDI problem for a network of unmanned vehicles does not have a solution for the decentralized architecture and vehicles need to exchange information. We tackle the FDI problem in a network of unmanned vehicles corresponding to centralized and semi-decentralized architectures in Sections 3.3.1 and 3.3.2, respectively. The necessary and sufficient conditions for solvability of the FDI problem are derived for centralized and semi-decentralized architectures. Our proposed FDI scheme is applied to the actuator FDI problem in the formation flight of satellites in Section 3.3.3. To show the applicability of our proposed structured residual set, two other case studies, namely, an F18-HARV (High Angle of Attack Research Vehicle) and a satellite with redundant reaction wheels are also presented in Sections 3.4 and 3.5, respectively.

In Chapter 4, a network of unmanned vehicles subject to large environmental disturbances is considered and a hybrid FDI scheme is developed to achieve robustness with respect to external disturbances. A hybrid architecture for a robust FDI is introduced in Section 4.2 that is composed of a bank of continuous-time residual generators and a discrete-event system (DES) fault diagnoser. Moreover, our proposed hybrid FDI algorithms are developed for both linear and nonlinear systems. Our developed hybrid scheme is applied to both a network of unmanned vehicles and an Almost-Lighter-Than-Air-Vehicle (ALTAV) in Sections 4.3 and 4.4, respectively.

In Chapter 5, we consider the effects of communication channels on fault detection and isolation. In Section 5.2, a communication channel is modeled by using the packet erasure model. By combining the vehicle dynamics and the channel packet erasure model a discrete-time Markovian jump model is obtained for the "entire" network. This motivate us to investigate the FDI problem for both the discrete-time and the continuous-time Markovian Jump Systems (MJS) in Sections 5.4 and 5.5, respectively and a novel geometric framework is introduced for solving these problems. We then apply our proposed algorithm to two case studies, namely, formation flight of satellites in the presence of imperfect communication links in Section 5.4.4 and VTOL (vertical take-off and landing) helicopter in Section 5.5.4.

In the final Chapter 6, we outline future directions of study in which more research is needed to be conducted.

The funding for much of the research described in this book was provided by grants from the Natural Sciences and Engineering Research Council (NSERC) of Canada, the Faculty of Engineering and Computer Science of Concordia University, and Tier-I Concordia University Research Chair (Khorasani).

Doha,Qatar *Nader Meskin*
Montreal, Canada *Khashayar Khorasani*

Acknowledgements

I would like to express deep gratitude to my academic advisor Prof. K. Khorasani for his consistent encouragement, motivation, guidance and support throughout my study at Concordia University. He has provided me with unquenchable enthusiasm, vision, and wisdom, which inspired me from the beginning to the end. I also wish to thank Dr. A. Aghdam for providing me the opportunity to study at Concordia University.

I would also like to thank all my friends and colleagues at Concordia University, particulary Dr. Amin Mannani, Dr. Behzad Samadi, Hani Khoshdel Nikkho, Ehsan Sobhani-Tehrani, Dr. Rasul Mohammadi, Dr. Amitabh Barua, Dr. Farzaneh Abdollahi, Dr. Elham Semsar-Kazerooni, and Dr. Mohammad El Bouyoucef, with whom I spent a lot of time both at work and in spare time. They made my four-year stay at Montreal enjoyable and memorable.

Last but by no means least, I am most sincerely grateful to my wife, Mojdeh, for her love and support over the years. She has always been a pillar of strength, inspiration and support. I could never thank my parents for their unconditional love and prayers, without which I would never have come so far.

Doha, Qatar

Nader Meskin

Acknowledgments

Contents

Acronyms

AHS	Automated Highway Systems
ALTAV	Almost-Lighter-Than-Air-Vehicle
AUV	Autonomous Underwater Vehicle
DES	Discrete-Event System
EFPRG	Extended Fundamental Problem in Residual Generation
ESA	European Space Agency
FDI	Fault Detection and Isolation
FPRG	Fundamental Problem in Residual Generation
HARV	High Angle of Attack Research Vehicle
HEFPRG	H_∞-based EFPRG
HFD	Hybrid Fault Diagnoser
LMI	Linear Matrix Inequality
lNFPRG	Locally Nonlinear FPRG
LTI	Linear Time-Invariant
MJS	Markovian Jump Systems
MJSD	Markovian Jump Systems with Time-Delay
NCS	Networked Control Systems
SFDIP	Structured Fault Detection and Isolation Problem
TPF	Terrestrial Planet Finder
VTOL	Vertical Take-Off and Landing
UAV	Unmanned Aerial Vehicle
UIO	Unknown Input Observer
u.o.s.	UnObservability Subspace

Chapter 1
Introduction

Recent years have witnessed a strong interest and intensive research activities in the area of networked autonomous unmanned vehicles, such as spacecraft formation flight, unmanned aerial vehicles, autonomous underwater vehicles, automated highway systems and multiple mobile robots. The envisaged networked architecture can provide surpassing performance capabilities and enhanced reliability; however, it requires extending the traditional theories of control, estimation, and Fault Detection and Isolation (FDI). Among the many challenges for these systems is development of autonomous cooperative control and estimation strategies that can maintain the group behavior and mission performance in presence of undesirable events such as failures and faults in the vehicles. In order to achieve this goal, the team should have the capability to detect and isolate vehicles faults and reconfigure the cooperative control algorithms to compensate for them. The main objective of this book is to explore the fault detection and isolation issues in networked multi-vehicle unmanned systems.

1.1 Statement of the Work

In this book, the problem of fault detection and isolation for networked multi-vehicle unmanned systems is addressed. These systems can be modeled as interconnected dynamical systems whose behavior depends not only on the individual vehicle dynamics, but also on the nature of their interactions. The fault detection and isolation (FDI) problem is among the most critical and central challenges in autonomous network of unmanned vehicles.

In this book, we first investigate the development of and comparison among three different FDI architectures for a network of unmanned vehicles, namely, centralized, decentralized and semi-decentralized topologies.

Each FDI architecture has its own advantages as well as challenges. In the centralized scheme, all the information is sent to a central FDI unit through communication network channels. Conceptually, the FDI algorithms that are used are similar to those for a single system and hence relatively simple and straightforward. However, the effects of communication channels such as data dropouts and time delays will lead to negative impacts on the performance of the FDI algorithms. In a decentralized scheme, all the information is processed locally and there is no central processing unit. In this scheme, the FDI problem should be solved locally at each vehicle on the basis of local observations, commands and information communicated from neighboring vehicles. Towards this end, the FDI solvability conditions are studied in this book for the centralized, the decentralized and the semi-decentralized architectures and comparisons are made among them.

One of the main challenges in the design of FDI algorithms is to distinguish the effects of disturbances from faults and to develop a robust FDI scheme without compromising the detection of incipient faults in the vehicles. In unmanned vehicles such UAV's, this problem is more challenging due to the vehicles' small size features and their higher sensitivity to disturbances such as wind gust. In this book, we investigate development and design of robust hybrid FDI approaches for both linear and nonlinear systems of networked unmanned vehicles.

Finally, the effects of communication channels on fault detection and isolation performance are investigated. A packet erasure channel model is considered for incorporating stochastic packet dropouts of communication channels. Combining the vehicle dynamics and the communication link yields a discrete-time Markovian Jump System (MJS) representation. This motivates the development of a geometric FDI framework for both the discrete-time and the continuous-time Markovian jump systems. The proposed MJS FDI strategies are then applied to formation flight of satellites as well as a Vertical Take-Off and Landing (VTOL) helicopter problem.

1.2 Literature Review

1.2.1 Fault Detection and Isolation (FDI)

Modern control systems are becoming increasingly more complex and issues of availability, cost efficiency, reliability, operating safety, and environmental protection concerns are gaining and receiving more attention. This requires a fault diagnosis system that is capable of detecting plant, actuator and sensor faults when they occur and of identifying and isolating the faulty component. A fault diagnosis algorithm consists of fault detection, isolation and identification steps. A traditional approach to fault diagnosis is based on the *hardware redundancy* [171] method which uses multiple sensors, actuators, computers and software to measure or control a particular value. Typically, a voting scheme is applied to the hardware redundant system to decide if and when a fault has occurred and its likely location among redundant components. The major problems encountered with hardware redundancy are the extra equipment, maintenance cost and the additional space required to accommodate the redundant components. An alternative approach for fault diagnosis is based on *analytical redundancy* which uses the redundant analytical relationships among system inputs and measured system outputs to generate residual signals where no extra hardware is required in this approach. In analytical redundancy schemes, the resulting difference that is generated from consistency checks of different variables is called a *residual* signal. Analytical redundancy schemes make use of a mathematical model of the monitored system and is often referred to as the *model-based approach* to fault diagnosis. Figure 1.1 illustrates the concepts of hardware and analytical redundancy [171].

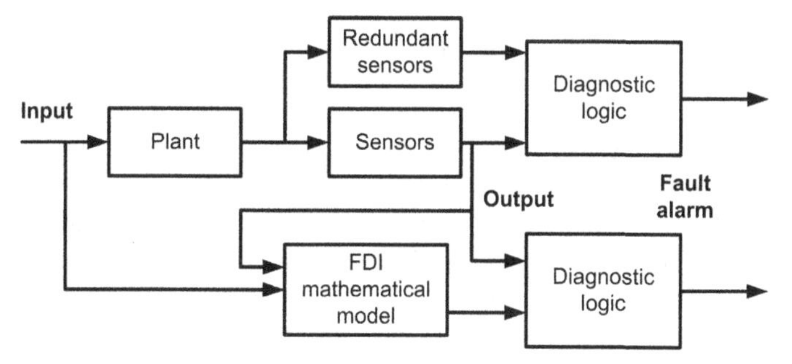

Fig. 1.1 Comparison between the hardware and the analytical redundancy based schemes for FDI [171].

The principle of model-based fault detection and isolation is depicted in Figure 1.2. Model-based residual generation techniques have been categorized by Patton et al. [149, 31, 150], Basseville and Nikiforov [12], and Gertler [76] as **a)** observer-based, **b)** parity equation, and **c)** identification and parameter estimation approaches.

In observer or filter-based approaches, the outputs of the system are estimated from the measurements (or subset of measurements) by using either Luenberger observers in the deterministic setting [13, 95, 119, 188, 58, 143, 142, 144, 42, 59, 34, 35, 33] or Kalman filters in a stochastic setting [124, 189, 11, 181, 12, 96, 20]. Subsequently, the weighted output estimation errors (or innovations in the stochastic case) is used as a residual. The flexibility in selecting observer gains has been fully utilized in the literature yielding a rich variety of FDI schemes. The increasing popularity of state-space models as well as the wide usage of observers in modern control theory and applications have made the observer-based FDI approach as one of the most common approaches in this domain. In parity space approach [41, 72, 32, 74, 76] resid-

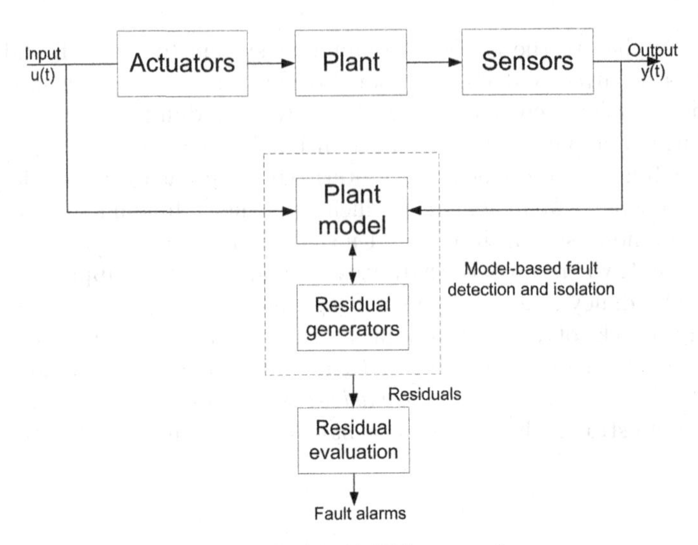

Fig. 1.2 The schematic for a model-based FDI approach.

ual signals (or parity vector) are generated based upon consistency checks on system input and output data over a given time window. It has been shown that some correspondence exists between the observer-based and the parity relation approaches [118, 149, 113, 151]. In other words, the parity relation approach is equivalent to the use of a dead-beat observer and a residual signal generated by a non-dead-beat observer is equivalent to a post-filtered residual generated by a dead-beat observer. This implies that the parity space method provides less design flexibility when compared with methods which are based on observers without any constrains [31].

Parameter estimation method [88, 89, 66] is developed based on system identification techniques. In this approach, the faults are reflected in the physical system parameters such as friction, mass, viscosity, resistance, inductance, capacitance, etc. The basic idea of the detection method is that the parameters of the actual process are estimated on-line by using the well-known parameter estimation schemes and the results are compared with the parameters of the reference model that is obtained initially under the fault-free condition. It should be pointed out that a typical limitation of the parameter estimation-based approaches is the fact that the input signal should be persistently exciting. This condition is satisfied if the input signal provides enough information to estimate the system parameters. However, many industrial systems may not allow feeding such persistently exciting signals as inputs to the process [190].

Upon the successful detection of a fault, the next step is to isolate a particular fault from others, i.e. fault isolation. While for fault detection a single residual signal is sufficient, a set of residuals is usually required for fault isolation. One way to fulfill the fault isolation task is to design a set of structured residuals. Each residual is designed to be sensitive to a subset of faults, whilst remaining residuals are to be insensitive to other faults. The residual set which has the required sensitivity to specific faults and insensitivity to other faults is known as the *structured residual set* [73]. The design procedure consists of two steps [31]. The first step is to specify the sensitivity and insensitivity relationships between residuals and faults according to the assigned isolation task. The second step is to design a set of residual generators according to the desired sensitivity and insensitivity relationships. In the second step different residual generation techniques such as observer-based or parity space approaches can be used for designing a residual set. The main advantage of the structured residual set is that the diagnostic analysis is simplified to determining which of the residuals have exceeded their thresholds. Several special residual schemes have been suggested in the literature [149, 120, 75]. In the dedicated residual set [120], all faults can be detected simultaneously, however it is difficult to design such residual sets for many practical systems, and generally there are no design degrees of freedom to achieve other desirable performance specifications and requirements such as robustness against uncertainty and modeling errors. More importantly, the necessary condition for designing such a residual set is *independence* of the fault signatures. In [149], a generalized residual set is introduced which can only detect a single fault although it has design degrees of freedom for achieving robustness against uncertainty and modeling errors. Figure 1.3 depicts the above two different structured residual sets for isolating three different faults $\{f_1, f_2, f_3\}$.

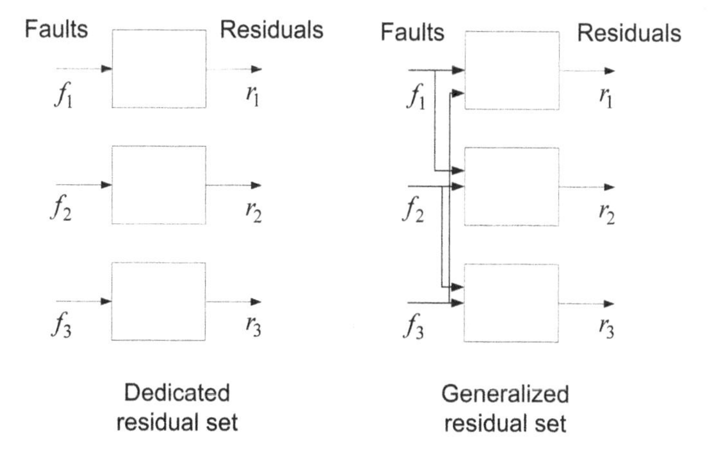

Fig. 1.3 The schematic of structured residual sets.

An alternative approach for achieving the isolability of faults is to design a *directional* residual vector [13, 95, 119, 188] which in response to a particular fault lies in a fixed and fault-specified direction in the residual space. With fixed directional residual vector, the fault isolation problem becomes that of determining which of the known fault signature directions the generated residual vector lies the closest to. The solvability conditions for generating a structured residual set are generally more relaxed when compared to the directional residual vector since in the later approach the design objective is to generate one residual vector with the above isolability condition, while in the former approach a set of residuals is generated and one may have more design degrees of freedom [119].

Recent years have witnessed a strong interest and considerable research activities on the design and analysis of FDI schemes for nonlinear systems [70, 102, 31, 190]. Most techniques in the literature constitute direct extensions of the approaches that are described above for linear systems. In [80], an extension of the parity space approach to nonlinear polynomial dynamic systems is proposed. Parity relation for a more general class of nonlinear systems was proposed in [101]. In [170], the concepts of the parity relation and parameter estimation fault detection techniques are combined for nonlinear systems. Many observer-based FDI approaches have also been developed in the literature for solving the nonlinear systems FDI problem [87, 1, 2]. The unknown input observer approach was extended to include nonlinear terms in [164, 166, 36]. Adaptive nonlinear observer based FDI scheme was proposed in [201, 91], and sliding mode observers have been designed in [175, 61, 92, 37, 200] for the nonlinear FDI problem. Persis and Isidori [153, 154] have extended Massoumnia's method [119, 120] to nonlinear systems. They showed that the problem of fault detection and isolation for nonlinear systems is solvable if and only if there is an unobservability distribution that leads to a quotient subsystem which is unaffected by all faults but one. In [155] the result of [154] is extended to a more realistic situation when measurement noise is present in the state affine system. Furthermore, the approach was extended to a special class of nonlinear systems. Hammouri et. al. [82] addressed the problem of fault detection and isolation for bilinear systems using the same geometric approach. In [126], the nonlinear geometric approach was successfully applied to a nonlinear longitudinal model of an aircraft subject to various types of actuator faults.

Model-based fault diagnosis approaches rely on the key assumption that a perfectly accurate and complete mathematical model of the system under supervision is available. However, such an assumption may not always be valid in practice. This problem has contributed to the rapid development of soft computing-based FDI methods [100]. Generally, the most popular soft computing techniques that are used in the FDI framework can be divided into three groups, namely [190]: **a)** neural networks [187, 69, 100, 146, 4, 173, 177, 178, 194, 110], **b)** Fuzzy logic-based techniques [100], and **c)** Evolutionary algorithms [116, 38, 176].

In this book, we adopt the model-based FDI approach according to the structured residual set concept along with the geometric FDI approach [120, 154] that is considered as the main methodology. In general, a primary advantage of geometric-type techniques is the formulation of the solutions in terms of very simple and intuitive concepts. The resulting consequence of this is that the problems are not being masked by heavy and complex mathematics as the solutions are easily reduced to matrix arithmetic as soon as one needs to compute them. Moreover, the FDI problem for linear and nonlinear systems can be tackled within a single unified framework.

1.2.2 Network of Unmanned Vehicles

Research in the cooperative control of unmanned vehicles is currently progressing in multiple directions and domains. Some examples are discussed in the paragraphs below.

Spacecraft Formation Flight: In recent years, there has been increased interest in research on the precision formation control of spacecraft [159, 160, 185, 14, 125, 114, 172, 157]. The use of multiple cooperating spacecraft has the potential to significantly expand the functionality, performance and reduce the overall operational costs. For instance, a formation of interferometric imaging spacecraft can achieve an optical imaging system with an aperture of kilometers long yielding the resolution that is required to image planets in the other solar systems. A number of space missions that are planned for the upcoming decades such as the NASA's terrestrial planet finder (TPF) [107] and the ESA's Darwin project [68] involve spacecraft formations.

Unmanned Aerial Vehicles: Advances in flight control technology have matured to a point where unmanned aerial vehicles (UAVs) are now routinely used in commercial and military applications. These advances have renewed extensive research interests in UAV cooperative control. Applications of this technology include coordinated rendezvous of UAVs [121, 16, 179], cooperative forest fire surveillance [28], cooperative search [15, 202, 93], and drag reduction via close formation flight [25, 40, 79, 191, 99].

Autonomous Underwater Vehicles: Over the past decade, numerous research activities have been conducted in the area of autonomous underwater vehicles (AUVs) where these vehicles are used in commercial, scientific and military applications [204]. Recently, researchers have also focused on AUV formations to accomplish more challenging tasks. Potential applications of AUV formations include oceanographic surveying at deep sea [21], operations under ice for exploration of Arctic areas and efficient monitoring of sub-sea oil rigs and installations [86].

Automated Highway Systems: In the past two decades, there has been an extensive research activity in automated highway systems (AHS) and new advances in computation and machine vision are enabling to bring these technologies closer to full implementation and utilization [18, 169].

Multiple Mobile Robots: Coordinated control of land mobile robots has been investigated in the past decade [53, 77, 141, 108]. Cooperative robots can be used to perform tasks that are too difficult for a single robot to perform alone. For example, a group of robots can be used to move large unusual objects [54, 105] , or to move a large number of objects [182]. Moreover, groups of robots can be used for terrain model acquisition [182], planetary exploration [106], or measuring radiation levels over a large area [5]. In [199], a group of robots are used for path obstruction.

While each of the above areas has its own unique challenges, several common themes can be found [64, 63]. In most cases, the vehicles are *dynamically decoupled* which implies that the motion of one does not directly affect the others. However, the vehicles are coupled through the mission goal that they are trying to accomplish. Each vehicle, through its sensors or communication links has limited information about the other vehicles which may be subjected to uncertainty and information transmission delays. The interactions among the vehicles motions make the network of vehicles an interconnected dynamic system whose behavior depends not only on the individual vehicles dynamics, but also on the nature of their interactions.

In recent years, only a few results on fault tolerant cooperative control of unmanned vehicle systems have been developed. In [145], a software architecture was developed to facilitate the fault tolerant cooperative control of teams of heterogenous mobile robots. In this architecture, if a robot fails, it cannot necessarily communicate its failure to its teammates and behavior-based fault detection approach was used to detect the failure in the robots. In this approach, each robot knows the tasks of other robots and if the specific task has not been completed in certain time interval, this will be interpreted as a failure in the corresponding robot.

A conversation protocol for failure detection is proposed in [39] where the robots can communicate with each other and send their faulty status to other robots. A fault tolerant algorithm is developed in [156] for the formation flight of UAVs in the presence of a failure in one or more of the UAVs. In this approach, it is assumed that the communication channel faults can be detected by the UAVs and an algorithm was proposed to reconfigure the formation according to the type of the fault.

In [43], a decentralized fault detection filter is designed as a combination of a game theoretic fault detection filter [42] with the decentralized filtering introduced in [174]. The approach is applied to a platoon of cars in an advanced vehicle control system. In [167], a decentralized detection filter for a large homogeneous collection of LTI systems is developed, with inspiration gained from platoons of vehicles with typical "look-ahead" structure. Recently in [50] a distributed, model-based, qualitative fault diagnosis scheme in developed for multi robot formations.

1.3 Features and Objectives of the Book

The main features and objectives of this book are as follows.

- **Fault detection and isolation (FDI) of systems with dependent fault signatures**
  Motivated by the FDI problem in a network of unmanned vehicles, we study systems with dependent fault signatures. For many classes of dynamical systems such as overactuated systems, fault signatures are generally dependent. This fault signature dependency may arise due to redundant actuators or coupling effects among sensors, actuators and plant faults. The first feature and contribution of this book is in the development, design, and analysis of fault detection and isolation schemes for systems with dependent fault signatures.

- **FDI for a network of unmanned vehicles with relative state measurements corresponding to centralized, decentralized and semi-decentralized architectures**
  Our proposed structured residual set is applied to the problem of actuator fault detection and isolation in a multi-vehicle system. Three different architectures, namely centralized, decentralized and semi-decentralized, are investigated for solving this problem and the necessary and sufficient solvability conditions for centralized and semi-decentralized architectures are derived based on the dynamics of each vehicle. Moreover, it is shown that the FDI problem does not have a solution for the decentralized architecture.

- **Robust hybrid FDI scheme for systems subject to large disturbances**
  A robust hybrid FDI scheme is developed for systems that are subject to large disturbances. A hybrid architecture is composed of a bank of continuous-time residual generators and a discrete-event system (DES) fault diagnoser. The main advantage of the proposed FDI scheme in comparison with previous work in the literature is that our algorithm is capable of distinguishing the effects of disturbances and incipient faults without imposing any limitations on the number of detectable faults in the system.

- **Fault detection and isolation of Markovian jump systems**
  A geometric FDI framework is developed for both discrete-time and continuous-time Markovian Jump Systems (MJS). A new geometric unobservable subspace property is derived for MJS systems and the notion of unobservability subspaces is introduced for these systems. The necessary and sufficient conditions for solvability of the fundamental problem of residual generation (FPRG) are derived for both discrete-time and continuous-time MJS systems. Furthermore, sufficient conditions for de-

signing H_∞-based FDI algorithms for MJS systems subject to input disturbances are also developed.

- **FDI for a network of unmanned vehicles in the presence of imperfect communication channels**
  By integrating a packet erasure channel model of each communication channel and the vehicle dynamics a discrete-time Markovian jump system representation is obtained. The proposed FDI algorithm for the discrete-time Markovian jump system is then applied for solving the FDI problem in a network of unmanned vehicles in presence of imperfect communication links

To summarize, the objectives of this book is to provide the first step towards the design of fault tolerant control systems for a network of unmanned vehicles by developing new fault detection and isolation schemes. We tackle different issues in these systems such as different FDI architectures, robustness to external disturbances and compensating the effects of communication channels in the framework of Markovian jump systems.

1.4 Outline of the Book

This book is organized as follows. In Chapter 2, a brief review of Structured Fault Detection and Isolation (SFDIP) for both linear and nonlinear systems are presented. A geometric FDI framework for solving the SFDIP problem is presented for linear and nonlinear systems in Sections 2.2 and 2.3, respectively.

In Chapter 3, we focus on the problem of fault detection and isolation in a network of unmanned vehicles without considering the effects of communication channels. It is shown in Section 3.1 that the actuator fault signatures in a network of unmanned vehicles with relative state measurements are dependent and the entire network can be considered as an overactuated system. This motivate us to develop a novel coding scheme that is applicable to systems with dependent fault signatures. Moreover, it is shown in Section 3.1.2 that the FDI problem for a network of unmanned vehicles corresponding to the decentralized architecture does not have a solution and vehicles need to exchange information for diagnosability. We also tackle the FDI problem in a network of unmanned vehicles corresponding to centralized and semi-decentralized architectures in Sections 3.3.1 and 3.3.2, respectively. The necessary and sufficient conditions for solvability of the FDI problem are derived for centralized and semi-decentralized architectures. The proposed FDI scheme is applied to the actuator FDI problem in the formation flight of satellites in Section 3.3.3. To demonstrate the applicability of the proposed structured residual set, two other case studies, namely, the first related to an F18-HARV and the second related to a satellite with redundant reaction wheels are also presented in Sections 3.4 and 3.5, respectively.

In Chapter 4, a network of unmanned vehicles that is subject to large environmental disturbances is considered and a hybrid FDI scheme is developed to achieve robustness with respect to external disturbances. A hybrid architecture for a robust FDI solution is introduced in Section 4.2 that is composed of a bank of continuous-time residual generators and a DES fault diagnoser. Moreover, hybrid FDI algorithms are then developed for both linear and nonlinear systems. developed hybrid scheme is applied to a network of unmanned vehicles and an Almost-Lighter-Than-Air-Vehicle (ALTAV) in Sections 4.3 and 4.4, respectively.

In Chapter 5, we consider the effects of communication channels on the solvability problem of fault detection and isolation. In Section 5.2, a communication channel is modeled by using the packet erasure model. By integrating the vehicle dynamics and the channel packet erasure model a discrete-time Markovian jump model is obtained for the "entire" network. This motivate us to investigate the FDI problem for both discrete-time and continuous-time Markovian Jump Systems (MJS) in Sections 5.4 and 5.5, respectively, by introducing a novel geometric framework for solving this problem. We then apply the proposed strategy to two case studies, namely, formation flight of satellites in the presence of imperfect communication links in Section 5.4.4 and VTOL (vertical take-off and landing) helicopter in Section 5.5.4. In the final Chapter 6, we outline future research directions in which further investigation is needed.

1.5 Notation

The following notation is used throughout this book. Script letters $\mathcal{X}, \mathcal{U}, \mathcal{Y}, ..$ denote real vector spaces. Matrices and linear maps are denoted by capital italic letters $A, B, C, ...$; the same symbol is used both for a matrix and its map; the zero space and zero vectors are denoted by 0. $\mathcal{B} = \text{Im } B$ denotes the image of B and Ker C denotes the kernel of C. The spectrum of A is denoted by $\sigma(A)$. A set Λ is said to be symmetric if $\lambda \in \Lambda$ implies $\lambda^* \in \Lambda$ where $*$ denotes the complex conjugate. For any positive integer k, $\mathbf{k}$ denotes the finite set $\{1, 2, \cdots, k\}$. If a map C is epic (onto), i.e. the matrix representation of C has full row rank, then C^{-r} denotes the right inverse of C (i.e., $CC^{-r} = I$). A subspace $\mathcal{S} \subseteq \mathcal{X}$ is termed A-invariant if $A\mathcal{S} \subseteq \mathcal{S}$. For an A-invariant subspace $\mathcal{S} \subseteq \mathcal{X}$, $A : \mathcal{S}$ denotes the restriction of A to $\mathcal{S}$, and $A : \mathcal{X}/\mathcal{S}$ denotes the map induced by A on the factor space $\mathcal{X}/\mathcal{S}$. For a linear system (C, A, B), $< \text{Ker } C|A >$ denotes the unobservable subspace of (C, A), $\otimes$ denotes the Kronecker product of matrices and for a given matrix A, A^N denotes $I_N \otimes A$ where I_N is an $N \times N$ identity matrix. For a given set $\mathbf{n}$, a *combination* is an un-ordered collection of the elements of $\mathbf{n}$ and is a subset of $\mathbf{n}$. The order of the elements in a combination is not important and the elements cannot be repeated. $C(n,k)$ denotes a number of k combinations (k-subset) of $\mathbf{n}$ which is equal to $\frac{n!}{k!(n-k)!}$. We denote by $||.||$ the standard norm in $\mathbb{R}^n$ and $\mathfrak{L}_2$ denotes the set of L_2 norm bounded signals. The symbol $*$ within a matrix represents the symmetric term of the matrix. For a given set N, $|N|$ denotes the cardinality of N.

A *directed graph* $\mathcal{G}$ consists of a set of vertices, or nodes, denoted by $\mathcal{V}$ and a set of arcs $\mathcal{A} \subset \mathcal{V}^2$, where $a = (v, w) \in \mathcal{A}$ and $v, w \in \mathcal{V}$. A directed graph is called weakly connected if there exist a path between every pair of distinct vertices ignoring the direction of the arcs.

Chapter 2
Geometric Approach to the Problem of Fault Detection and Isolation (FDI)

In this chapter, we briefly review a structured fault detection and isolation problem (SFDIP) for both linear and nonlinear systems based on geometric approaches that are developed in [120] and [154].

This chapter is organized as follows. In Section 2.1 the Structured Fault Detection and Isolation Problem (SFDIP) is presented. The geometric FDI framework for linear and nonlinear systems are then reviewed in Sections 2.2 and 2.3, respectively. In 2.4, common actuator fault modes that are considered in this book are presented.

2.1 Structured Fault Detection and Isolation (FDI)

In this section, the structured fault detection and isolation (FDI) problems for both linear and nonlinear systems are reviewed. As pointed out in Chapter 1, the structured residual set is one of the common way for accomplishing the fault isolation task and it has more design degrees of freedom with respect to other approaches such as the directional residual vector scheme. The design procedure consists of two steps [31], the first step is to specify the sensitivity and insensitivity relationships between the residuals and the faults according to the assigned isolation task, and the second step is to design a set of residual generators according to the desired sensitivity and insensitivity relationships. In the second step different residual generation techniques such as the observer-based or the parity space approach can be used for designing a residual set. In this book, we adopt the geometric FDI approach for designing the residual generators which will be discussed in Sections 2.2 and 2.3.

Consider the following linear time-invariant system

$$\begin{aligned}\dot{x}(t) &= Ax(t) + Bu(t) + \sum_{i=1}^{k} L_i m_i(t) \\ y(t) &= Cx(t)\end{aligned} \tag{2.1}$$

where $x \in \mathcal{X}$ is the state of the system with dimension n and the nonlinear system

$$\begin{aligned}\dot{x}(t) &= f(x(t)) + g(x(t))u(t) + \sum_{i=1}^{k} l_i(x(t))m_i(t) \\ y(t) &= h(x(t))\end{aligned} \tag{2.2}$$

where state x is defined in a neighborhood X of the origin in $\mathbb{R}^n$, $u \in \mathcal{U}, y \in \mathcal{Y}$ are input and output signals with dimensions m and q, respectively, $m_i \in \mathcal{M}_i$ are fault modes with dimension k_i and L_i's ($l_i(x)$) are fault signatures .

The fault modes together with the fault signatures may be used to model the effects of actuator faults, sensor faults (for linear systems only) and system faults on the dynamics of the system. For modeling a fault in the i-th actuator, L_i ($l_i(x)$) is chosen as the i-th column of matrix B ($g(x)$) and the fault mode m_i is chosen to model the type of a fault. For example a complete failure of an actuator can be represented and modeled by $m_i = -u_i$.

A system fault can be represented by a potential variation in the parameters of the matrix A for linear system (2.1) as shown below:

$$\begin{aligned}\dot{x}(t) &= (A + \Delta A)x(t) + Bu(t) \\ y(t) &= Cx(t)\end{aligned} \tag{2.3}$$

As an example, a change in the i-th row and the j-th column element of the matrix A can be modeled according to [31]:

$$\Delta Ax(t) = I_i \Delta a_{ij} x_j(t).$$

where x_j is the j-th element of the vector x and I_i is the n-dimensional vector with all zero elements except a 1 in the i-th element. Defining a signal $m_i(t) \triangleq \Delta a_{ij} x_j(t)$ as an external input and the fault signature $L_i = I_i$, equation (2.3) can then be rewritten as:

$$\begin{aligned}\dot{x}(t) &= Ax(t) + Bu(t) + L_i m_i(t) \\ y(t) &= Cx(t)\end{aligned}$$

Similarly, a system fault for the nonlinear system (2.2) can be represented by a potential variation in the parameters of the vector field $f(x)$.

It should be noted that sensor faults can initially be modeled as additive inputs in the measurement equation

$$y(t) = Cx(t) + \sum_{j=1}^{q} E_j n_j(t) \tag{2.4}$$

where E_j is an $q \times 1$ unit vector with a one at the j-th position and $n_j \in \mathbb{R}$ is a sensor fault mode, which correspond to a fault in the j-th sensor. For example a complete failure of the j-th sensor can be represented and modeled by $n_j(t) = -c_j x(t)$ where c_j is the j-th row of the matrix C.

The sensor fault signature can also be modeled for linear systems as an input to the system [42, 120, 85]. Following [42], let f_j be the solution to $E_j = C f_j$. The new states can be defined according to $\bar{x} = x + \sum_{j=1}^{q} f_j n_j$, where the state space representation for the new states can be written as

$$\begin{aligned} \dot{\bar{x}}(t) &= A\bar{x}(t) + Bu(t) + \sum_{j=1}^{q} L_j m_j(t) \\ y(t) &= C\bar{x}(t) \end{aligned} \tag{2.5}$$

where $L_j = \begin{bmatrix} f_j \ A f_j \end{bmatrix}$ and $m_j(t) = \begin{bmatrix} \dot{n}_j(t) \ -n_j(t) \end{bmatrix}^{\top}$.

The Structured Fault Detection and Isolation Problem (SFDIP) as introduced in [120], [73] is defined formally as design of a dynamic residual generator that takes the observations $u(t)$ and $y(t)$ as inputs and generates a set of residual signals $r_i(t), i \in \mathbf{p}$ with the following properties.

1. When no fault is present, all the residuals $r_i(t)$ decay asymptotically to zero.
2. The residuals $r_i(t)$ for $i \in \Omega_j$ are affected by a fault of the j-th component, and the other residuals $r_\alpha(t)$ for $\alpha \in \mathbf{p} - \Omega_j$ are decoupled from this fault.

The prespecified family of *coding sets* $\Omega_j \subseteq \mathbf{p}, j \in \mathbf{k}$ should be chosen such that, by knowing which of the $r_i(t)$ are zero and which are not, one can uniquely identify the fault. It is evident that we should have $\cup_{j=1}^{k} \Omega_j = \mathbf{p}$, i.e. each residual is at least affected by one fault. The resulting residual set which has the corresponding required sensitivity to specific faults and insensitivity to other faults is known as the *structured residual set* [73]. For detecting all possible faults in the system, no coding set should be empty. The minimum requirement for fault isolation is that all the coding sets are distinct. Coding sets satisfying these two requirements are defined as *weakly isolating* .

Definition 2.1 ([73]) *The weakly isolating coding sets $\Omega_i, i \in \mathbf{k}$ are defined as being* strongly isolating *if for each $i, j \in \mathbf{k}$, $i \neq j$*

$$\Omega_i \nsubseteq \Omega_j \tag{2.6}$$

A strongly isolating coding set prevents incorrect fault detection when some of the residuals in Ω_i do not exceed the respective thresholds while the others do. For the given coding sets $\Omega_i, i \in \mathbf{k}$, the finite set $\Gamma_j \subset \mathbf{k}, j \in \mathbf{p}$ is defined as the collection of all $i \in \mathbf{k}$ for which the i-th failure mode affects the j-th residual, i.e. $\Gamma_j = \{i \in \mathbf{k} | j \in \Omega_i\}$.

In the Extended Fundamental Problem of Residual Generation (EFPRG) introduced in [120] for linear systems and in the nonlinear fundamental problem in residual generation (lNFPRG) introduced in [154] for nonlinear system, the coding set of $\Omega_i = \{i\}$ was chosen. This coding set is also called a *dedicated residual set* [31] which is inspired by the dedicated observer scheme that was proposed by Clark [44] (Figure 1.3). In this coding scheme, one looks to design a set of filters which generates k residuals $r_i(t), i \in \mathbf{k}$ such that a fault in the i-th component $m_i(t) \neq 0$ can only affect the residual $r_i(t)$ and no other residuals $r_j(t)(i \neq j)$. With this coding scheme one can detect and isolate all faults simultaneously. Another commonly used scheme in designing the residual set is to make each residual sensitive to all but one fault which is known as a *generalized residual set* [149]. By utilizing this residual set one cannot simultaneously detect and isolate faults in two or more channels.

Once the residual signals are generated according to given coding sets $\Omega_i, i \in \mathbf{k}$, the final step for performing fault detection and isolation is to determine the threshold values J_{th_i} and the evaluation functions $J_{r_i}(t)$. Various evaluation functions have been introduced in [56] as follows:

- Instantaneous value of the residual signal, i.e. $J_{r_i} = r_i$.
- The average value of the residual signal over a time interval $[t-T, t]$, i.e.

$$J_{r_i} = ||r||_a = \frac{1}{T} \int_{t-T}^{t} r(\tau) d\tau \tag{2.7}$$

- Root-mean-square (RMS) which measures the average energy over a time interval $(0, T)$, i.e.,

$$J_{r_i} = \left(\frac{1}{T} \int_{0}^{T} ||r(\tau)||^2 d\tau \right)^{1/2} \tag{2.8}$$

- Truncated RMS which measures the average energy over a time interval $[t-T, t]$, i.e

$$J_{r_i} = \frac{1}{T} \int_{t-T}^{t} ||r(\tau)||^2 d\tau \tag{2.9}$$

The threshold value for each of the above residual evaluation functions can be selected as

$$J_{th_i} = \sup_{m_i=0, d \in \mathfrak{D}} J_{r_i} \tag{2.10}$$

where d represents the input disturbance or measurement noise that is applied and present in the models (2.1) and (2.2), and $\mathfrak{D}$ denotes the set of allowable disturbances, for instance, $\mathfrak{D} = \mathfrak{L}_2$ or $\mathfrak{D}$ can be selected as a set of Gaussian white noise. Based on the thresholds and the evaluation functions, the occurrence of a fault can then be detected and isolated by using the following decision logic

$$J_{r_j}(t) > J_{th_j}, \quad \forall j \in \Omega_i \Longrightarrow m_i \neq 0, \quad i \in \mathbf{k} \tag{2.11}$$

2.2 Geometric Approach to FDI of Linear Systems

In this section the geometric approach that was introduced in [120] for the SFDIP of linear systems is reviewed. In general, a primary advantage of geometric-type techniques is the formulation of the results in terms of simple and intuitive concepts that avoid heavy and complex mathematics as they are represented in terms of the matrix arithmetic that can be easily computed. In the geometric approach to the fault detection and isolation problem certain subspaces play a central role as defined below.

Definition 2.2 ([10]) *A subspace $\mathcal{W}$ is a (C, A)-invariant subspaces (conditioned invariant) if $A(\mathcal{W} \cap Ker\ C) \subseteq \mathcal{W}$.*

It is simple to show that $\mathcal{W}$ is (C, A)-invariant if and only if there exists a map $D : \mathcal{Y} \to \mathcal{X}$ such that $(A + DC)\mathcal{W} \subseteq \mathcal{W}$. The set of all (C, A)-invariant subspaces containing a given subspace $\mathcal{L} \subseteq \mathcal{X}$ is denoted by $\mathfrak{W}(A, C, \mathcal{L})$. It can be shown that $\mathfrak{W}(A, C, \mathcal{L})$ is closed under intersection and is nonempty ($\mathcal{X} \in \mathfrak{W}(A, C, \mathcal{L})$); therefore, it admits an infimum, the *minimal (C, A)-invariant containing* $\mathcal{L}$, which will be denoted by $\mathcal{W}^* = \inf\ \mathfrak{W}(A, C, \mathcal{L})$. The following algorithm [10] can be used for finding $\mathcal{W}^*$.

Algorithm 2.1 ([10]) *Subspace $\mathcal{W}^*$ coincides with the last term of the sequence*

$$\begin{aligned} \mathcal{Z}_0 &= \mathcal{L} \\ \mathcal{Z}_i &= \mathcal{L} + A(\mathcal{Z}_{i-1} \cap\ Ker\ C), \quad (i \in \boldsymbol{k}) \end{aligned}$$

where the value of $k \leq n$ is determined by the condition $\mathcal{Z}_{k+1} = \mathcal{Z}_k$.

The dual of the (C, A)-invariant subspace is the (A, B)-invariant (controlled invariant), which is defined as follows:

Definition 2.3 ([10, 192]) *A subspace $\mathcal{V} \subseteq \mathcal{X}$ is said to be (A, B)-invariant (controlled invariant) if*

$$A\mathcal{V} \subseteq \mathcal{V} + Im\ B \tag{2.12}$$

Similarly, the set of all (A, B)-invariants contained in a given subspace $\mathcal{E}$ (denoted by $\mathfrak{V}(A, B, \mathcal{E})$) is closed under the subspace addition operator and nonempty, hence, it admits a supermum, which is denoted by $\mathcal{V}^* = \max \mathfrak{V}(A, B, \mathcal{E})$. The following algorithm [10] can be used for finding $\mathcal{V}^*$.

Algorithm 2.2 ([10]) *Subspace $\mathcal{V}^*$ coincides with the last term of the sequence*

$$\mathcal{Z}_0 = \mathcal{E}$$
$$\mathcal{Z}_i = \mathcal{E} \cap A^{-1}(\mathcal{Z}_{i-1} + ImB), \quad (i \in \mathbf{k})$$

where the value of $k \leq n$ is determined by the condition $\mathcal{Z}_{k+1} = \mathcal{Z}_k$.

Next, some specific subclass of (C, A)-invariants is introduced.

Definition 2.4 ([10]) *Let $\mathcal{S}$ be a (C, A)-invariant subspace containing a subspace $\mathcal{L} \subseteq \mathcal{X}$, $\mathcal{S}$ is said to be self-hidden with respect to $\mathcal{L}$ if*

$$\mathcal{S} \subseteq \mathcal{W}^* + \ Ker\ C \tag{2.13}$$

where $\mathcal{W}^* = \inf\ \mathfrak{W}(A, C, \mathcal{L})$.

It can be shown that the class $\Psi_{(C,\mathcal{L})}$ of (C, A)-invariants self-hidden with respect to $\mathcal{L}$, i.e.

$$\Psi_{(C,\mathcal{L})} = \{\mathcal{S}\ :\ A(\mathcal{S} \cap \text{Ker } C) \subseteq \mathcal{S}, \mathcal{L} \subseteq \mathcal{S}, \mathcal{S} \subseteq \mathcal{W}^* + \text{Ker } C\} \tag{2.14}$$

is closed under the subspace addition, and hence it has the supremum element which can be found by the following lemma.

Lemma 2.1 ([10]) *The supremum of $\Psi_{(C,\mathcal{L})}$ is*

$$\mathcal{W}^* + \mathcal{V}_2^* \tag{2.15}$$

where $\mathcal{V}_2^* = \max \mathfrak{V}(A, \mathcal{L}, Ker\ C)$.

In the *geometrical approach* to the fault detection and isolation problem certain unobservability subspaces play a central role [119, 120] as defined below.

Definition 2.5 *A subspace $\mathcal{S}$ is a (C, A)* unobservability subspace *(u.o.s.) [119] if $\mathcal{S} =< Ker\ HC|A + DC >$ for some output injection map $D : \mathcal{Y} \to \mathcal{X}$ and measurement mixing map $H : \mathcal{Y} \to \mathcal{Y}$.*

It can be shown [119] that for an unobservability subspace $\mathcal{S}$,

$$\mathcal{S} =< \mathcal{S} + \text{Ker } C|A + DC > \tag{2.16}$$

The next theorem provides an alternative characterization of the u.o.s. which is independent of the maps D and H (this is the dual to Theorem 5.3 in [192] for controllability subspaces).

Theorem 2.1 *Let $\mathcal{S} \subseteq \mathcal{X}$ and define the family $\mathcal{G}_{(A,C)}$ as follows*

$$\mathcal{G}_{(A,C)} = \{\mathscr{S} : \mathscr{S} = \mathcal{S} + A^{-1}\mathscr{S} \cap\ Ker\ C\} \tag{2.17}$$

$\mathcal{S}$ is an (C, A) u.o.s. if and only if

$$A(\mathcal{S} \cap \; Ker\, C) \subseteq \mathcal{S} \quad (\mathcal{S} \textit{ is conditioned invariant}) \tag{2.18}$$

and

$$\mathcal{S} = \mathscr{S}^* \tag{2.19}$$

where $\mathscr{S}^*$ is the maximal element of $\mathcal{G}_{(A,C)}$.

The maximal element $\mathscr{S}^*$ can be computed by invoking the following algorithm

$$\mathscr{S}^0 = \mathcal{X}; \quad \mathscr{S}^\mu = \mathcal{S} + A^{-1}\mathscr{S}^{\mu-1} \cap \text{ Ker } C, \quad \mu \in \mathbf{n} \tag{2.20}$$

Given an u.o.s. $\mathcal{S}$, a measurement mixing map H can be computed from $\mathcal{S}$ by solving the equation $\text{Ker}HC = \text{Ker}C + \mathcal{S}$. Let $\underline{D}(\mathcal{S})$ denote the class of all maps $D : \mathcal{Y} \to \mathcal{X}$ such that $(A + DC)\mathcal{S} \subseteq \mathcal{S}$. The notation $\mathfrak{S}(A, C, \mathcal{L})$ refers to the class of u.o.s. containing $\mathcal{L} \subseteq \mathcal{X}$. The class of u.o.s. is closed under intersection; therefore, it contains an infimal element $\mathcal{S}^* = \inf \mathfrak{S}(A, C, \mathcal{L})$. Moreover,

$$\mathcal{S}^* =< \text{Ker } C + \mathcal{W}^* | A + DC > \tag{2.21}$$

The following lemma shows the equivalency between the infimal unobservability subspace $\mathcal{S}^*$ containing a given subspace $\mathcal{L}$ and the supremum element of class $\Psi_{(C,\mathcal{L})}$.

Lemma 2.2 ([10]) *The infimal (C, A)-unobservability subspace $\mathcal{S}^*$ containing $\mathcal{L}$ is the supremum (C, A)-invariant self-hidden with respect to $\mathcal{L}$, i.e. $\mathcal{S}^* = \mathcal{W}^* + \mathcal{V}_2^*$.*

The above lemma provides a convenient approach for finding $\mathcal{S}^*$ by using the Algorithms 2.1 and 2.2. In [119] an alternative algorithm for computing $\mathcal{S}^*$ is proposed as follows.

Algorithm 2.3 ([119]) *The subspace $\mathcal{S}^*$ coincides with the last term of the sequence*

$$\begin{aligned} \mathcal{Z}_0 &= \mathcal{X} \\ \mathcal{Z}_i &= \mathcal{W}^* + (A^{-1}\mathcal{Z}_{k-1}) \cap Ker\, C \quad (i \in \boldsymbol{k}) \end{aligned}$$

where the value of $k \leq n$ is determined by the condition $\mathcal{Z}_{k+1} = \mathcal{Z}_k$.

The next theorem provides the necessary and sufficient conditions for solvability of EFPRG problem ($\Omega_i = \{i\}$) based on the concept of unobservability subspaces.

Theorem 2.2 ([120]) *EFPRG problem has a solution if and only if there exist (C, A) unobservability subspaces*

$$\mathcal{S}_i^* = \inf \mathfrak{S}(A, C, \sum_{j=1, j \neq i}^{k} \mathcal{L}_j), \quad i \in \boldsymbol{k} \tag{2.22}$$

such that $\mathcal{S}_i^* \cap \mathcal{L}_i = 0$, $i \in \boldsymbol{k}$.

The major step in generating the residual r_i is to incorporate the image of the fault signatures that requires to decouple $(L_j (j \neq i))$ in the unobservability subspace of r_i and then factor out the unobservable subspace in a manner that in the remaining factor space those faults do not appear. The associated necessary condition for this purpose states that the image of L_i should not intersect with the unobservable subspace of r_i, so that a fault in the i-th component is manifested in the residual r_i.

According to [120], let $\mathcal{S}_i^*$ denote an u.o.s. that satisfies Theorem 2.2, then there exists a map $D_0 \in \underline{D}(\mathcal{S}_i^*)$ and H_i such that $\mathcal{S}_i^* =< \text{Ker } H_i C | A + D_0 C >$, where H_i is a solution to Ker $H_i C = \mathcal{S}_i^* +$Ker C. Let M_i be a unique solution of $M_i P_i = H_i C$ and $A_0 = (A + D_0 C : \mathcal{X}/\mathcal{S}_i^*)$, P_i is the canonical projection of $\mathcal{X}$ on $\mathcal{X}/\mathcal{S}_i^*$ and $(A + D_0 C : \mathcal{X}/\mathcal{S}_i^*)$ denotes an induced map of $A + D_0 C$ on the factor space $\mathcal{X}/\mathcal{S}_i^*$ which satisfies the following equation

$$P_i(A + D_0 C) = (A + D_0 C : \mathcal{X}/\mathcal{S}_i^*) P_i \tag{2.23}$$

By construction, the pair (M_i, A_0) is observable, hence there exists a D_1 such that $\sigma(F_i) = \Lambda$, where $F_i = A_0 + D_1 M_i$ and Λ is an arbitrary symmetric set. Let $D = D_0 + P_i^{-r} D_1 H_i$, $E_i = P_i D$ and $G_i = P_i B$. The following detection filter generates the desired residual which is only affected by L_i and is decoupled from other faults,

$$\begin{aligned} \dot{w}_i(t) &= F_i w_i(t) - E_i y(t) + G_i u(t) \\ r_i(t) &= M_i w_i(t) - H_i y(t) \end{aligned} \tag{2.24}$$

Definition 2.6 ([192]) *Let $p = (p_1, ..., p_N) \in \mathbb{R}^N$ and consider polynomials $\phi_i(\alpha_1, \dots, \alpha_N), i \in \boldsymbol{k}$ with coefficients in $\mathbb{R}$. A* variety *$V \subseteq \mathbb{R}^N$ is defined as the locus of common zeros of a finite number of polynomials $\phi_1, ..., \phi_k$. V is* proper *if $V \neq \mathbb{R}^N$. Let $A, B, ...$ denote matrices with elements in $\mathbb{R}$ and suppose $\prod(A, B, ...)$ is some property which may be asserted about them. Let V be a proper variety. We say that $\prod$ is* generic relative *to V provided the property $\prod$ does not hold only for points $p \in V$, and $\prod$ is* generic *provided that such a V exists. In other words, the set of points on which the generic property does not hold has a Lebesgue measure of zero.*

The generic solvability conditions for the EFPRG problem are stated as follows [120].

Proposition 2.1 ([120]) *Let A, C and L_i denote arbitrary matrices of dimensions $n \times n$, $q \times n$ and $n \times k_i$, respectively, and $v = \sum_{i=1}^{k} k_i$. The EFPRG problem generically has a solution if and only if*

$$v \leq n \tag{2.25}$$

and

$$v - \min\{k_i, i = 1, ..., k\} < q \tag{2.26}$$

In the next theorem, the solvability condition for the SFDIP problem is presented. For given coding sets $\Omega_j, j \in \mathbf{k}$, the finite sets $\Gamma_i, i \in \mathbf{p}$ are defined as the collection of all $j \in \mathbf{k}$ for which the j-th failure mode affects the i-th residual, i.e. $\Gamma_i = \{j \in \mathbf{k} | i \in \Omega_j\}$. It follows that the sets Γ_i contain all the information corresponding to the coding sets $\Omega_j, j \in \mathbf{k}$.

Theorem 2.3 *For a given family of coding sets, the SFDIP problem has a solution if and only if*

$$\mathcal{S}^*_{\Gamma_i} \cap \mathcal{L}_j = 0, \quad j \in \Gamma_i, \quad i \in \mathbf{p} \tag{2.27}$$

where

$$\mathcal{S}^*_{\Gamma_i} = \inf \mathfrak{S}(A, C, \sum_{j \notin \Gamma_i} \mathcal{L}_j), \qquad i \in \mathbf{p} \tag{2.28}$$

Proof: According to Theorem 2.2, the SFDIP problem can be solved as p separate EFPRG problems. Each residual $r_i(t), i \in \mathbf{p}$ can be generated by applying the EFPRG results to the following model

$$\begin{aligned}
\dot{x}(t) &= Ax(t) + Bu(t) + \bar{L}_1 \bar{m}_1(t) + \bar{L}_2 \bar{m}_2(t) \\
y(t) &= Cx(t)
\end{aligned}$$

where $\bar{L}_1 = \{L_j | j \in \Gamma_i\}$, $\bar{L}_2 = \{L_j | j \in \mathbf{k} - \Gamma_i\}$, $\bar{m}_1 = \{m_j | i \in \Gamma_i\}$ and $\bar{m}_2 = \{m_j | i \in \mathbf{k} - \Gamma_i\}$. Hence, according to Theorem 2.2 in order to decouple the residual $r_i(t)$ from $\bar{m}_2(t)$, there should exists an unobservability subspace that contains all fault signature in $\bar{L}_2$, i.e. $\mathcal{S}^*_{\Gamma_i} = \inf \mathfrak{S}(A, C, \sum_{j \notin \Gamma_i} \mathcal{L}_j)$. Moreover, the necessary condition for manifesting the effects of faults in components $m_j, j \in \Gamma_i$ in the residual r_i is that $\mathcal{S}^*_{\Gamma_i}$ does not have intersection with $L_j, j \in \Gamma_i$, i.e. condition (2.27). ■

A family of fault signatures satisfying the conditions of Theorem 2.2 is designated as a *strongly detectable* family. It follows from Theorem 2.2 that a necessary condition for existence of a solution to the EFPRG problem is that

$$\text{Rank}\{[L_1, L_2, ..., L_k]\} = \sum_{i=1}^{k} k_i \tag{2.29}$$

which implies that there should be no dependency among the fault signatures. The following example demonstrates situations in which the family of fault signatures is not *strongly detectable*.

Example 2.1. Consider the following overactuated system

$$\begin{aligned} \dot{x}(t) &= Ax(t) + Bu(t) + \sum_{i}^{4} B_i m_i(t) \\ y(t) &= Cx(t) \end{aligned} \tag{2.30}$$

where

$$A = \begin{bmatrix} 1 & 1 & 1 \\ 0 & -1 & 2 \\ 2 & 0 & -4 \end{bmatrix}, B = \begin{bmatrix} 1&0&1&1 \\ 0&1&1&0 \\ 0&1&0&1 \end{bmatrix}, C = \begin{bmatrix} 0&1&0 \\ 0&0&1 \\ 1&0&2 \end{bmatrix}$$

This system has 4 inputs such that rank $B = 3$. Therefore, the family of actuator fault signatures for this system is clearly not strongly detectable.

Example 2.2. Consider the following system

$$\dot{x}(t) = Ax(t) + Bu(t) \tag{2.31}$$

$$y(t) = Cx(t) + \sum_{j=1}^{3} E_j \mu_j \tag{2.32}$$

where

$$A = \begin{bmatrix} 1 & 2 & 1 & 0 & 0 \\ 0 & 1 & 3 & -1 & 2 \\ 3 & -1 & 2 & 0 & 1 \\ 0 & 1 & 1 & 1 & 1 \\ 0 & 0 & 2 & 1 & 1 \end{bmatrix}, B = \begin{bmatrix} 1&0 \\ 1&0 \\ 0&1 \\ 0&1 \\ 0&0 \end{bmatrix}, C = \begin{bmatrix} 1&1&0&0&1 \\ 0&1&1&0&0 \\ 0&0&1&1&0 \end{bmatrix}$$

As mentioned earlier, sensor faults can be modeled as input faults to the system. Using the state space coordinate transformation $\bar{x} = x + \sum_{j=1}^{3} f_j \mu_j$, the state space representation for the new state is written according to equation (2.5) where

$$L_1 = \begin{bmatrix} 1&1 \\ 0&0 \\ 0&3 \\ 0&0 \\ 0&0 \end{bmatrix}, L_2 = \begin{bmatrix} -2 & 1 \\ 2 & -1 \\ -1 & -10 \\ 0 & 1 \\ 0 & -2 \end{bmatrix}, L_3 = \begin{bmatrix} 0 & 0 \\ 0 & -1 \\ 0 & 0 \\ 1 & 1 \\ 0 & 1 \end{bmatrix} \tag{2.33}$$

If one considers the actuator faults L_4 and L_5 as the first and second columns of B, it can be easily checked that the family of fault signatures $L_1, ..., L_5$ is not strongly detectable. Therefore, one cannot detect all faults simultaneously in this system.

2.3 Geometric Approach to FDI of Nonlinear Systems

In this section, the geometric FDI approach introduced in [154] for nonlinear systems is briefly reviewed. This scheme provides us with the necessary and sufficient conditions for solving the problem of generating a structured residual set for general nonlinear systems. It is assumed that the nonlinear system is described by the following model

$$\begin{aligned} \dot{x}(t) &= f(x(t)) + g(x(t))u + \sum_{i=1}^{k} l_i(x(t))m_i(t) \\ y(t) &= h(x(t)) \end{aligned} \tag{2.34}$$

with the state x defined in a neighborhood X of the origin in $\mathbb{R}^n$; the input is denoted by $u \in \mathbb{R}^m$; the output is denoted by $y \in \mathbb{R}^q$; the fault modes are denoted by $m_i \in \mathbb{R}^{k_i}$, $l_i(x)$'s are the fault signatures, $f(x)$, $g(x)$ and $l_i(x)$'s are smooth vector fields, $h(x)$ is a nonlinear smooth mapping, and $f(0) = 0, h(0) = 0$. It is assumed that $span\{l_i^1(x), ..., l_i^{k_i}(x)\}, i \in \mathbf{k}$ is nonsingular where l_i^j denotes the j-th column of l_i.

Definition 2.7 ([90]) *The distribution Δ is said to be* conditioned invariant *(or (h, f) invariant) for system (2.34) if it satisfies*

$$[g_i, \Delta \cap Ker\{dh\}] \subseteq \Delta, \quad for\ all\ i = 0, ..., m \tag{2.35}$$

where $g_0(x) = f(x)$, $Ker\{dh\}$ is the distribution annihilating the differentials of the rows of the mapping $h(x)$ and $[f, g]$ denotes the Lie derivative of f and g.

For a given distribution $\mathcal{L}$, the following algorithm is proposed in [153] to determine the smallest condition invariant distribution which contains $\mathcal{L}$ (denoted by $\Sigma_*^{\mathcal{L}}$):

Algorithm 2.4 ([153]) *Consider the following sequence of distributions*

$$\begin{aligned} \mathcal{S}_0 &= \bar{\mathcal{L}} \\ \mathcal{S}_{k+1} &= \bar{\mathcal{S}}_k + \sum_{i=0}^{m} [g_i, \bar{\mathcal{S}}_k \cap Ker\{dh\}] \end{aligned}$$

where $\bar{\mathcal{S}}$ denotes the involutive closure of $\mathcal{S}$. Suppose there is an integer k^* such that $\mathcal{S}_{k^*+1} = \bar{\mathcal{S}}_{k^*}$, then $\Sigma_*^{\mathcal{L}} = \bar{\mathcal{S}}_{k^*}$ and $\Sigma_*^{\mathcal{L}}$ is involutive and is the smallest conditioned invariant containing $\mathcal{L}$.

According to [153], it is more convenient to work with a dual object, i.e. with a codistribution as defined next.

Definition 2.8 ([154]) *A codistribution Π is said to be conditioned invariant if*

$$L_{g_i}\Pi \subseteq \Pi + span\{dh\}, \quad for\ all\ i = 0, ..., m \tag{2.36}$$

where $span\{dh\}$ is the codistribution spanned by the differentials of the rows of the mapping $h(x)$.

It is shown in [90] that if $\Delta \cap \mathrm{Ker}\{dh\}$ is a smooth distribution and $\Pi = \Delta^\perp$ is a smooth codistribution, then Π satisfies (2.36). In [153], the following algorithm was proposed for defining an observability codistribution associated with system (2.34).

Algorithm 2.5 ([153]) *Consider the system (2.34) and let Θ denote a fixed codistribution. The observability codistribution algorithm (o.c.a.) that characterizes this codistribution is given according to the following procedure:*

$$Q_0 = \Theta \cap span\{dh\}$$

$$Q_{k+1} = \Theta \cap \big(\sum_{i=0}^{m} L_{g_i} Q_k + span\{dh\}\big)$$

Suppose all codistributions of the sequences are nonsingular, so that there is an integer $k^* \leq n-1$ such that $Q_k = Q_{k^*}$ for all $k > k^*$ and set $\Pi^* = Q_{k^*}$. The following notation is then used to stress the dependency of Π^* on Θ:

$$\Pi^* = o.c.a.(\Theta) \tag{2.37}$$

Definition 2.9 *The codistribution Π is an* observability codistribution *for (2.34) if it satisfies (2.36) and*

$$o.c.a.(\Pi) = \Pi \tag{2.38}$$

If Θ is a conditioned invariant codistribution, then o.c.a.(Θ) is the maximal observability codistribution contained in Θ. The next theorem which is proposed in [153] states one of the important properties of the observability codistribution.

Theorem 2.4 ([153]) *Consider system (2.34) with $m_i = 0$. Let Π be an observability codistribution such that $\Pi = span\{\Phi_1\}$ where $\Phi_1 : U^\circ \to \mathbb{R}^{n_1}$ and U° is a neighborhood of fixed point x°. One can then find a local state diffeomorphism at x° and a local output diffeomorphism at $y^\circ = h(x^\circ)$ such that in the new local coordinates the system (2.34) is described by the following equations*

$$\begin{aligned}
\dot{z}_1 &= f_1(z_1, z_2) + g_1(z_1, z_2)u \\
\dot{z}_2 &= f_2(z_1, z_2, z_3) + g_2(z_1, z_2, z_3)u \\
\dot{z}_3 &= f_3(z_1, z_2, z_3) + g_3(z_1, z_2, z_3)u \\
y_1 &= h(z_1) \\
y_2 &= z_2
\end{aligned} \tag{2.39}$$

where $z_1 = \Phi_1(x)$*, and in the new coordinates any vector field* $l(x)$ *in* $\Pi^{\perp}$ *can be expressed in the form*

$$\left(0 \; l_2^{\top}(z_1, z_2, z_3) \; l_3^{\top}(z_1, z_2, z_3)\right)^{\top} \tag{2.40}$$

The observability codistribution is a special type of condition invariant codistribution where the z_1-subsystem, in which z_2 can be replaced by y_2 and viewed as an independent "input", namely system

$$\begin{aligned} \dot{z}_1 &= f_1(z_1, y_2) + g_1(z_1, y_2)u \\ y_1 &= h_1(z_1) \end{aligned} \tag{2.41}$$

satisfies the observability rank condition and where by imposing certain mild assumptions [153], one can design an asymptotic observer for the state z_1.

In the nonlinear fundamental problem in residual generation (lNFPRG) that is introduced in [154], the family of the coding sets $\Omega_i = \{i\}$ was chosen. The solvability of the lNFPRG problem is stated as follows.

Theorem 2.5 ([154]) *The lNFPRG has a solution if and only if there exist observability codistributions*

$$\Pi_i^* = o.c.a.((\Sigma_*^{\mathcal{L}_i})^{\perp}) \tag{2.42}$$

such that

$$(span\{l_i\})^{\perp} + \Pi_i^* = T^*X \tag{2.43}$$

where $\mathcal{L}_i = span\{l_1(x), ..., l_{i-1}(x), l_{i+1}(x), ..., l_m(x)\}$*.*

If such observability codistribution exist, then by using Theorem 2.4, the z_1-subsystem corresponding to each Π_i^* is described in the new coordinates by

$$\begin{aligned} \dot{z}_1 &= f_1(z_1, y_2) + g_{i1}(z_1, y_2)u_i + l_{i1}(z_1, y_2)m_i \\ y_1 &= h_1(z_1) \end{aligned} \tag{2.44}$$

Consequently, it is now possible to design an observer for the above subsystem which generates the state estimate $\hat{z}_1$. In [154], one way to design an observer for the z_1-subsystem (2.44) is presented based on a result from Gauthier-Kupca [71]. Once the observer is designed, the residual signal r_i can be generated according to

$$r_i = y_1 - h_1(\hat{z}_1) \tag{2.45}$$

The next theorem provides the necessary and sufficient conditions for the solvability of the SFDIP problem for given coding sets Ω_i's for the nonlinear system (2.34).

Theorem 2.6 *The SFDIP problem for the nonlinear system* (2.34) *has a solution for given coding sets Ω_i's if and only if there exist observability codistributions*

$$\Pi^*_{\Gamma_i} = o.c.a.((\Sigma_*^{\mathcal{L}_{\Gamma_i}})^\perp), \quad i \in \mathbf{p} \tag{2.46}$$

such that

$$(span\{l_j\})^\perp + \Pi^*_i = T^*X, \quad \forall j \in \Gamma_i, \quad i \in \mathbf{p} \tag{2.47}$$

where $\mathcal{L}_{\Gamma_i} = span\{l_j(x) | j \notin \Gamma_i\}$, $i \in \mathbf{p}$.

Proof: According to Theorem 2.5, the SFDIP problem can be solved as p separate lNFPRG problems. Each residual $r_i(t), i \in \mathbf{p}$ can be generated by applying the EFPRG results to the following model

$$\begin{aligned} \dot{x} &= f(x) + g(x)u + \bar{L}_1(x)\bar{m}_1(t) + \bar{L}_2(x)\bar{m}_2(t) \\ y &= h(x) \end{aligned}$$

where $\bar{L}_1 = \{l_i(x) | i \in \Gamma_i\}$, $\bar{L}_2 = \{l_i(x) | i \in \mathbf{k} - \Gamma_i\}$, $\bar{m}_1 = \{m_i | i \in \Gamma_i\}$ and $\bar{m}_2 = \{m_i | i \in \mathbf{k} - \Gamma_i\}$. The residual $r_i(t)$ should be decoupled from all faults in the components $\bar{m}_2$ and should be affected by faults in the components $m_j, j \in \Gamma_i$. Hence, according to Theorem 2.5, there should exist an observability codistribution that contains all fault signatures in $\bar{L}_2$, i.e. $\Pi^*_{\Gamma_i} =$ o.c.a.$((\Sigma_*^{\mathcal{L}_{\Gamma_i}})^\perp)$. Moreover, the necessary condition for manifesting the effects of faults in the components $m_j, j \in \Gamma_i$ is (2.47). ■

2.4 Actuator Fault Modes

Actuator faults that are considered in this book for simulation studies include [22]: (i) freezing or lock in-place (LIP) fault, (ii) float fault, (iii) hard-over fault (HOF), and (iv) loss of effectiveness (LOE) fault. In case of the LIP fault, the actuator states freeze at a particular value and will not respond to the subsequent commands. The HOF is characterized by the actuator moving to its upper or lower saturation limits regardless of the commanded signal. The actuator transient response time is bounded by its rate limits. Float fault occurs when the actuator floats with zero output and does not contribute to the control authority. Loss of effectiveness fault is characterized and represented by lowering the actuator gain with respect to its nominal value. The various types of actuator faults discussed above are mathematically parameterized as follows [22]

$$u_{true} = \begin{cases} u_{cmd} & \text{No Fault} \\ k(t)u_{cmd} & 0 < \varepsilon \le k(t) < 1, \forall t \ge t_F \text{ (LOE)} \\ 0 & \forall t \ge t_F \text{ (Float)} \\ u_{cmd}(t_F) & \forall t \ge t_F \text{ (LIP)} \\ u_m \quad \text{or} \quad u_M & \forall t \ge t_F \text{ (HOF)} \end{cases} \tag{2.48}$$

where t_F denotes the time of fault occurrence in the actuator, k denotes the loss of effectiveness coefficient where $k \in [\varepsilon, 1]$ and ε denotes the minimum loss of effectiveness, and u_m and u_M denote the minimum and the maximum values of the input, respectively. The above cases can be integrated into a single general representation as follows, $u_{true} = \sigma k u_{cmd} + (1 - \sigma)\bar{u}$, where u_{true} is the actuator output, u_{cmd} is the output of the controller (which is actually an input to the actuator), $\sigma = 1$ and $k = 1$ correspond to the no fault operating regime, $\sigma = 1$ and $\varepsilon < k < 1$ correspond to the loss of effectiveness fault, and $\sigma = 0$ corresponds to other types of fault scenarios, namely Float, LIP and HOF. Finally, $u_m \le \bar{u} \le u_M$ denotes the state at which the actuator is locked for the float, lock-in-place, and hard-over fault cases.

It should be emphasized that all the algorithms that are developed throughout this book do not actually depend on the above fault modeling, namely the above actuator faults are only used in the simulation studies.

Chapter 3
FDI in a Network of Unmanned Vehicles: Ideal Communication Channels

In this chapter, we address the problem of actuator fault detection and isolation of a network of unmanned vehicles corresponding to three different architectures, namely centralized, decentralized, and semi-decentralized. It is shown that the fault signatures in a network of unmanned vehicles with relative state measurements are dependent and the overall system is overactuated. This motivates us to investigate the development, design, and analysis of a fault detection and isolation (FDI) scheme for both linear and nonlinear systems with dependent fault signatures. Earlier versions of the work presented in this chapter have partially appeared in [127, 128, 129, 131].

This chapter is organized as follows. In Section 3.1, the actuator fault detection and isolation problem in a network of unmanned vehicles is formalized associated with different architectures, namely centralized, decentralized and semi-decentralized. In Section 3.2, new coding sets are introduced for fault detection and isolation of both linear and nonlinear systems with dependent fault signatures. In Section 3.3, the proposed FDI algorithm is applied to the actuator fault detection and isolation of a network of unmanned vehicles. Simulation results for the FDI of formation flight of satellites are presented in Section 3.3.3. Moreover, two other case studies, namely, actuator FDI problem for the F18-HARV and FDI of a satellite with redundant reaction wheel configuration are considered in Sections 3.4 and 3.5, respectively.

3.1 FDI Problem Formulation in a Network of Unmanned Vehicles

In this section, the problem of actuator fault detection and isolation in a network of N homogenous (for sake of simplicity) vehicles is formalized. It is assumed that each vehicle dynamics is governed by the following linear model and representation

$$\dot{x}_i(t) = Ax_i(t) + Bu_i(t) + \sum_{k=1}^{a} L_k m_{ik}(t) \tag{3.1}$$

where $x_i \in \mathbb{R}^n$ is the state of the i-th vehicle, $u_i \in \mathbb{R}^a$ is the input signal of the i-th vehicle and the fault signature L_k represents a fault in the k-th actuator of a vehicle, i.e., L_k is the k-th column of the matrix B. It is assumed that the matrix B is full column rank. Each vehicle has the following relative state measurements

$$z_{ij}(t) = C(x_i(t) - x_j(t)) \qquad j \in \mathrm{N}_i \tag{3.2}$$

where the set $\mathrm{N}_i \subseteq \mathbf{N} \setminus i$ represents the set of vehicles that vehicle i can sense ($\mathbf{N} = \{1, ..., N\}$) and is designated as the neighboring set of vehicle i, and $z_{ij} \in \mathcal{Z}_i, j \in \mathrm{N}_i$ represents the state measurement relative to the other vehicles.

It is assumed that the pair (A, C) is observable. Let $\mathrm{N}_i = \{i_1, i_2, ..., i_{|\mathrm{N}_i|}\}$. We have $z_i(t) = [z_{ii_1}^\top(t), z_{ii_2}^\top(t), \cdots, z_{ii_{|\mathrm{N}_i|}}^\top(t)]^\top$, and equation (3.2) can be rewritten as

$$z_i(t) = C_i x(t) \tag{3.3}$$

where $x(t) = [x_1^\top(t), x_2^\top(t), \cdots, x_N^\top(t)]^\top$ and C_i can be constructed from the neighboring set N_i and matrix C. The vehicles and their neighborhood sets together form a directed graph, where each node represents a vehicle and an arc leads from node i to node j if $j \in \mathrm{N}_i$. It is assumed that this graph is weakly connected, i.e. there exists a path between every pair of distinct vertices ignoring the direction of the arcs.

In the following discussions we will formulate three different architectures for FDI of actuator faults in the above network of unmanned vehicles. Namely, we are interested in centralized, decentralized and semi-decentralized configurations as described and defined below. It will be shown that actuator fault signatures in the centralized and semi-decentralized architectures are dependent, and therefore new coding schemes need to be developed for systems with dependent fault signatures (Section 3.2). Moreover, we will show that the actuator FDI problem does not have a solution in the decentralized architecture.

In this chapter, it is assumed that the communication links among the vehicles and the FDI units are ideal, i.e. there exist no networked-induced delay, packet dropout or quantization errors due to the communication links. This assumption will be relaxed subsequently in Chapter 5.

3.1.1 Centralized Architecture

In the centralized architecture, all the information should be sent to a central FDI unit through the "entire" communication network channels. The overall system can be modeled equivalently as follows

$$\begin{aligned}\dot{x}(t) &= A^N x(t) + B^N u(t) + \sum_{k=1}^{N}\sum_{j=1}^{a} \bar{L}_{kj} m_{kj}(t) \\ z(t) &= \bar{C}x(t)\end{aligned} \tag{3.4}$$

where $A^N = I_N \otimes A$, $B^N = I_N \otimes B$, $\otimes$ denotes the Kronecker product, I_N is an $N \times N$ identity matrix, $\bar{L}_{kj}$ is the $(k-1) \times a + j$-th column of B^N, and

$$u = \begin{bmatrix} u_1 \\ \vdots \\ u_N \end{bmatrix}, z = \begin{bmatrix} z_1 \\ \vdots \\ z_N \end{bmatrix}, \bar{C} = \begin{bmatrix} C_1 \\ \vdots \\ C_N \end{bmatrix} \tag{3.5}$$

It follows that the entire state x is not fully observable from the relative state measurements z and states of the centroid of the vehicles cannot be determined from z. Since the pair $(A^N, \bar{C})$ in not observable, one can first determine the observable subspace of system (3.4) as

$$\begin{aligned}\dot{x}^O(t) &= A^{N-1}x^O(t) + \bar{B}^O u(t) + \sum_{k=1}^{N}\sum_{j=1}^{a} \bar{L}^O_{kj} m_{kj}(t) \\ \bar{z}(t) &= C^{N-1}x^O(t)\end{aligned} \tag{3.6}$$

where

$$x^O(t) = \begin{bmatrix} x_1(t) - x_2(t) \\ x_1(t) - x_3(t) \\ \vdots \\ x_1(t) - x_N(t) \end{bmatrix}, \bar{B}^O = \begin{bmatrix} B & -B & 0 & \cdots & 0 \\ B & 0 & -B & \cdots & 0 \\ \vdots & & \ddots & & \vdots \\ B & 0 & 0 & \cdots & -B \end{bmatrix}_{n(N-1)\times aN} \tag{3.7}$$

and $C^{N-1} = I_{N-1} \otimes C$, $\bar{L}^O_{kj}$ is the $(k-1) \times a + j$-th column of $\bar{B}^O$, and $\bar{z}(t) = Tz(t)$, where T is an output transformation that specifies the relative states with respect to the first vehicle. Such a transformation exists since it is assumed that the graph of the network is weakly connected. Without loss of generality, the relative states between vehicle one and other vehicles are taken as observable states of the entire network. It can be verified that the column rank of $\bar{B}^O$ is $a \times (N-1)$ and system (3.6) is overactuated.

3.1.2 Decentralized Architecture

In this section, the decentralized architecture is considered where there exist no communication links among the vehicles and we wish to determine whether each vehicle can detect and isolate its own faults by using only local signals, namely u_i and z_i. Since the output measurement z_i depends on the state of the neighboring vehicles, the following nodal model should be considered for the i-th vehicle for designing a decentralized FDI filter, namely

$$\begin{aligned}\dot{x}_{\mathrm{N}_i}(t) =& A^{|\mathrm{N}_i|+1}x_{\mathrm{N}_i}(t) + B^{|\mathrm{N}_i|+1}u_{\mathrm{N}_i}(t) + \sum_{j=1}^{a}\bar{L}_{1j}m_{ij}(t) \\ &+ \sum_{k=1}^{|\mathrm{N}_i|}\sum_{j=1}^{a}\bar{L}_{(k+1)j}m_{i_kj}(t) \\ z_i(t) =& \bar{C}_{\mathrm{N}_i}x_{\mathrm{N}_i}(t)\end{aligned} \tag{3.8}$$

where

$$x_{\mathrm{N}_i}(t) = \begin{bmatrix} x_i(t) \\ x_{i_1}(t) \\ \vdots \\ x_{i_{|\mathrm{N}_i|}}(t)\end{bmatrix}, u_{\mathrm{N}_i}(t) = \begin{bmatrix} u_i(t) \\ u_{i_1}(t) \\ \vdots \\ u_{i_{|\mathrm{N}_i|}}(t)\end{bmatrix}, z_i(t) = \begin{bmatrix} z_{ii_1}(t) \\ z_{ii_2}(t) \\ \vdots \\ z_{ii_{|\mathrm{N}_i|}}(t)\end{bmatrix} \tag{3.9}$$

and $\bar{L}_{(k+1)j}$ is the $k \times a + j$-th column of $B^{|\mathrm{N}_i|+1}$. It should be noted that $\bar{L}_{1j}$'s represent the fault signatures of the i-th vehicle in the nodal system (3.8). It follows that the entire state x_{N_i} is not fully observable from the relative state measurements z_i, and the states of the centroid of vehicle i and its neighbors cannot be determined from z_i. Since the pair $(A^{|\mathrm{N}_i|+1}, \bar{C}_{\mathrm{N}_i})$ is not observable, one can first try to obtain the observable part of system (3.8) as given below

$$\begin{aligned}\dot{x}^O_{\mathrm{N}_i}(t) &= A^{|\mathrm{N}_i|}x^O_{\mathrm{N}_i}(t) + \bar{B}^O_{\mathrm{N}_i}u_{\mathrm{N}_i}(t) + \sum_{j=1}^{a}\bar{L}^O_{1j}m_{ij}(t) + \sum_{k=1}^{|\mathrm{N}_i|}\sum_{j=1}^{a}\bar{L}^O_{(k+1)j}m_{i_kj}(t) \\ z_i(t) &= C^{|\mathrm{N}_i|}x^O_{\mathrm{N}_i}(t)\end{aligned} \tag{3.10}$$

where

$$x^O_{\mathrm{N}_i}(t) = \begin{bmatrix} x_i(t) - x_{i_1}(t) \\ x_i(t) - x_{i_2}(t) \\ \vdots \\ x_i(t) - x_{i_{|\mathrm{N}_i|}}(t) \end{bmatrix}, \quad \bar{B}^O_{\mathrm{N}_i} = \begin{bmatrix} B & -B & 0 & \cdots & 0 \\ B & 0 & -B & \cdots & 0 \\ \vdots & & \ddots & & \vdots \\ B & 0 & 0 & \cdots & -B \end{bmatrix}_{n(|N_i|)\times a(|N_i|+1)} \tag{3.11}$$

and $\bar{L}^O_{(k+1)j}$ is the $k \times a + j$-th column of $\bar{B}^O_{\mathrm{N}_i}$. Let us define matrices $\bar{B}^O_i$ and $\bar{B}^O_{i_k}$ as in $\bar{B}^O_{\mathrm{N}_i} = [\bar{B}^O_i, \bar{B}^O_{i_1}, \cdots, \bar{B}^O_{i_{\mathrm{N}_i}}]$. We are now in the position to introduce our main theorem in this subsection.

Theorem 3.1 *The decentralized FDI problem is not solvable for a network of unmanned vehicles governed by* (3.8) *by using only the relative state measurements.*

Proof: In order to generate the residual signals $r_{ij}, j \in \mathbf{a}$ for the i-th vehicle which is only affected by the fault signal m_{ij}, using only the signals u_i and z_i, the residual signals should be decoupled from faults in the neighboring vehicles $m_{i_k l}(t)$, $k = 1, ..., |N_i|$, $l \in \mathbf{a}$, input signals $u_j, j \neq i$ and m_{il}, $l \in \mathbf{a}, l \neq j$. However, since the fault signatures $\bar{L}^O_{(k+1)j}$'s are the columns of $\bar{B}^O_{\mathrm{N}_i}$, the term $\sum_{k=1}^{|\mathrm{N}_i|} \bar{\mathcal{B}}^O_{i_k}$ includes both fault signatures and input channel matrices of other vehicles. Therefore, by invoking the result of Theorem 2.3, the existence of the $(C^{|\mathrm{N}_i|}, A^{|\mathrm{N}_i|})$ unobservability subspaces $\mathcal{S}^*_j = \inf \underline{\mathcal{S}}(\sum_{k=1}^{|\mathrm{N}_i|} \bar{\mathcal{B}}^O_{i_k} + \sum_{k=1,k\neq j}^{a} \bar{\mathcal{L}}^O_{1k})$ such that $\mathcal{S}^*_j \cap \bar{\mathcal{L}}_{1j} = 0, j \in \mathbf{a}$, is the necessary and sufficient condition for generating the residual signal $r_{ij}(t)$. This residual signal is only affected by m_{ij} and is decoupled from all other fault signatures of the i-th vehicle and $\sum_{k=1}^{|\mathrm{N}_i|} \bar{\mathcal{B}}^O_{i_k}$. However, it follows from equation (3.11) that $\bar{\mathcal{L}}_{1j} \subset \sum_{k=1}^{|\mathrm{N}_i|} \bar{\mathcal{B}}^O_{i_k}, j \in \mathbf{a}$ and $\bar{\mathcal{L}}_{1j} \subseteq \mathcal{S}^*_j, j \in \mathbf{a}$. Therefore, it can be concluded that the above decentralized FDI problem does not have a solution for a network of unmanned vehicles by using only the relative state measurements, and each vehicle cannot even detect its own fault by using the local information u_i and z_i. It can be shown similarly that it is not also possible to generate the generalized structured residual set for system (3.8). ∎

3.1.3 Semi-decentralized Architecture

In the semi-decentralized architecture, it is assumed that local communication links exist between each vehicle and its neighbors and the control signals u_i are communicated among them. The problem is to determine how each vehicle detects and isolates not only its own actuator faults but also actuator faults of its neighbors by using the signals u_{N_i} and z_i. The observable nodal model (3.10) is considered for the i-th vehicle.

It can be verified that the column rank of $\bar{B}^O_{\mathrm{N}_i}$ is $a \times |\mathrm{N}_i|$ and the fault signatures $\bar{L}^O_{kj}$'s are not strongly detectable. This is due to the fact that one needs to control $|\mathrm{N}_i|$ vehicles in the nodal system (3.10) in order to control the relative states of $|\mathrm{N}_i| + 1$ vehicles. In other words, actuator redundancy is present and the system can be categorized as an overactuated system.

As shown above, in centralized and semi-decentralized architectures, the fault signatures are dependent and hence concurrent faults in all actuators cannot be detected and isolated. This provides us with the motivation to determine and investigate a suitable coding set for a family of dependent fault signatures. In the next section, we propose the structured residual set for both linear and nonlinear systems with dependent fault signatures.

3.2 Proposed Structured Fault Detection and Isolation Scheme

For many class of dynamical systems such as overactuated systems, fault signatures are generally dependent. Generally speaking, fault signatures are called *dependent* if an effect of one fault can be written as a linear combination of effects of other faults. This dependency of fault signatures may arise due to redundant actuators or coupling effects among sensor, actuator and plant faults. In this section, we investigate the development, design, and analysis of a fault detection and isolation scheme for both linear and nonlinear systems with dependent fault signatures. Due to dependency among fault signatures, the dedicated residual set cannot be utilized and consequently all faults cannot be detected simultaneously. To determine a suitable residual set, an isolability index μ is first introduced for a family of fault signatures as the maximum number of concurrent faults that can be detected and isolated. Next, two structured residual sets are proposed and developed for achieving a specific isolability index. The geometric approaches of [120] and [154] are used for generating these new residual sets for linear and nonlinear systems, respectively.

Definition 3.1 *For a family of fault signatures $L_1, \cdots, L_k$ for linear system (2.1) ($l_1(x), \cdots, l_k(x)$ for nonlinear system (2.34)), the maximum value of $\mu \leq k$ where one can detect and isolate the occurrence of up to μ concurrent faults is denoted as the isolability index .*

According to the above definition, Theorem 2.2 shows that the isolability index of a family of k fault signatures is k if and only if the family is *strongly detectable* .

The next theorem provides the necessary and sufficient conditions for a given coding sets that can be used for detecting and isolating up to μ concurrent faults in a given family of fault signatures $\mathcal{FL} = \{L_1, L_2, ..., L_k\}$ for the linear system (2.1). The same statement can be derived for the nonlinear system (2.34) by substituting $l_i(x)$'s with L_i's.

Theorem 3.2 *Consider the weakly isolating coding sets $\Omega_i, i \in \mathbf{k}$ and a given family of fault signatures $\mathcal{FL} = \{L_1, L_2, ..., L_k\}$ for linear system* (2.1). *Let for each l combination $L_{i_1}, ..., L_{i_l}$ of L_i's, $\Omega_{i_1 i_2 ... i_l} = \bigcup_{j=i_1}^{i_l} \Omega_j$. The family of fault signatures $\mathcal{FL}$ has an isolability index of μ if and only if the SFDIP problem has a solution for the coding sets $\Omega_i, i \in \mathbf{k}$ with the following property that for each two different combinations $L_{i_1}, ..., L_{i_l}$ and $L_{j_1}, ..., L_{j_h}$ of L_i's, $l, h \in \{1, ..., \mu\}$, we have*

$$\Omega_{i_1 i_2 ... i_l} \neq \Omega_{j_1 j_2 ... j_h} \tag{3.12}$$

Proof: (only if) Let a family of fault signatures $L_1, ..., L_k$ has the isolability index of μ, then for any occurrence of $l \leq \mu$ concurrent faults, there should exists a unique set of residuals that is affected by these faults. According to the definition of the coding sets, a set of residuals that is affected by l faults $L_{i_1}, ..., L_{i_l}$ is $\Omega_{i_1 i_2 ... i_l}$. Therefore, the uniqueness of $\Omega_{i_1 i_2 ... i_l}$ leads to the equation (3.12).

(if) It is trivial to observe that if the SFDIP problem has a solution for the coding sets $\Omega_i, i \in \mathbf{k}$ with the above property, then one can detect and isolate up to μ concurrent faults. ■

It should be emphasized that for $\mu = k$, there exists only one k-combination of L_i's, namely $L_1, ..., L_k$, and hence condition (3.12) implies that for having an isolability index of $\mu = k$, we should have

$$\Omega_{i_1 i_2 ... i_l} \neq \Omega_{1...k}$$

for any $l < k$ where $\Omega_{1...k} = \bigcup_{j=1}^{k} \Omega_j$.

Remark 3.1. It can be shown that the coding sets of the EFPRG problem [120] ($\Omega_i = \{i\}$) and the coding sets of the generalized residual set [149] ($\Omega_i = \mathbf{k} - \{i\}$) satisfy the necessary and sufficient conditions of the above theorem for $\mu = k$ and $\mu = 1$, respectively.

As mentioned in Section 2.1, in order to prevent incorrect fault detection and isolation, the coding sets Ω_i should be strongly isolating. The same criteria can be considered for the coding sets $\Omega_{i_1 i_2 ... i_l}$.

Definition 3.2 *The coding sets that satisfy the conditions of Theorem 3.2 are said to be strongly isolating with index μ if for each two different $1 \leq l \leq \mu$ combination $L_{i_1}, ..., L_{i_l}$ and $L_{j_1}, ..., L_{j_l}$ of L_i's*

$$\Omega_{i_1 i_2 ... i_l} \nsubseteq \Omega_{j_1 j_2 ... j_l}, \quad \textbf{and} \quad \Omega_{i_1 i_2 ... i_l} \nsupseteq \Omega_{j_1 j_2 ... j_l} \tag{3.13}$$

Moreover, if the SFDIP problem for a given family of fault signatures has a solution for a strongly isolating coding set with index μ, then we call the isolability index of that family as a strongly isolability index μ.

Lemma 3.1 *The strong isolability index μ of a given family of fault signatures is either $\mu = k$ or $\mu < k - 1$.*

Proof: Let $\Omega_i, i \in \mathbf{k}$ be strongly isolating with index $\mu = k-1$. Consider that concurrent faults have occurred in all fault signatures. Then, all residuals will be affected by these concurrent faults since we have $\cup_{j=1}^{k} \Omega_j = \mathbf{p}$. Moreover, since the strong isolability of fault signatures is μ, it follows that for each $k-1$ combination $L_{i_1}, ..., L_{i_{k-1}}$ of L_i's, we have $\Omega_{i_1 i_2 \ldots i_{k-1}} \neq \Omega_{1,\ldots,k} = \bigcup_{j=1}^{k} \Omega_j$. Therefore, one can also detect that there exist concurrent faults in all channels. Moreover, by assumption one can detect and isolate up to $k-1$ concurrent faults. Hence $\mu = k$. ■

It should be emphasized that not every coding set that satisfies Theorem 3.2 is strongly isolating with index μ. For instance, consider the coding sets $\Omega_1 = \{1, 2\}$, $\Omega_2 = \{3, 4\}$, and $\Omega_3 = \{2, 3\}$. It can easily be verified that these coding sets satisfy the conditions of Theorem 3.2 with $\mu = 2$ but they are not strongly isolating with index 2. Indeed it follows that $\Omega_{2,3} \subset \Omega_{1,2}$.

Since strong isolability index is more desirable and it prevents incorrect detection and isolation of faults, we will focus on strong isolability index. Recall from Section 2.1 that for the given coding sets $\Omega_i, i \in \mathbf{k}$, the finite set $\Gamma_j \subset \mathbf{k}, j \in \mathbf{p}$ is defined as the collection of all $i \in \mathbf{k}$ for which the i-th failure mode affects the j-th residual, i.e. $\Gamma_j = \{i \in \mathbf{k} | j \in \Omega_i\}$.The next theorem illustrates how one can construct coding sets that have the strong isolability index $\mu < k - 1$.

Theorem 3.3 *Let $\Gamma_j, j \in \mathbf{p} = \{1, ..., p\}$ be defined as the $k - \mu$ combinations of the set $\mathbf{k}$ where $p = C(k, k - \mu)$. Then the corresponding sets $\Omega_i, i \in \mathbf{k}$ defined as*

$$\Omega_i = \{j \in \mathbf{p} | i \in \Gamma_j\} \tag{3.14}$$

are strongly isolating with index μ.

Proof: Consider two different combinations $i_1, ..., i_l$ and $j_1, ..., j_h$ of the set $\mathbf{k}$, $l, h \in \{1, ..., \mu\}$. In order to show that equation (3.12) holds it is sufficient to show that

$$\mathbf{p} - \Omega_{i_1 i_2 \ldots i_l} \neq \mathbf{p} - \Omega_{j_1 j_2 \cdots j_h} \tag{3.15}$$

where

$$\mathbf{p} - \Omega_{i_1 i_2 \ldots i_l} = \{j \in \mathbf{p} | i_k \notin \Gamma_j, \ \forall i_k \in \{i_1, ..., i_l\}\} \tag{3.16}$$

is the set of residuals that are not affected by faults $m_{i_1}, ..., m_{i_l}$. Assume that $l \leq h$. Since two combinations $i_1, ..., i_l$ and $j_1, ..., j_h$ are different, there exists $j_t \in \{j_1, ..., j_h\}$ such that $j_t \in \mathbf{k} - \{i_1, i_2, ..., i_l\}$. Since the sets $\Gamma_i, i \in \mathbf{p}$ are defined as $k - \mu$ combinations of the set $\mathbf{k}$ and $l \leq \mu$, there exists a combination Γ_j such that $j_t \in \Gamma_j$ and $i_k \notin \Gamma_j$, $\forall i_k \in \{i_1, ..., i_l\}$. Therefore, we have

$$\left.\begin{array}{c} j_t \in \{j_1, ..., j_h\} \\ j_t \in \Gamma_j \end{array}\right\} \Rightarrow j \notin \mathbf{p} - \Omega_{j_1 j_2 ... j_h}$$

$$i_k \notin \Gamma_j,\ \forall i_k \in \{i_1, ..., i_l\} \Rightarrow j \in \mathbf{p} - \Omega_{i_1 i_2 ... i_l}$$

which shows that equation (3.15) holds. The combination Γ_j can be found by first selecting j_t and then selecting $k - \mu - 1$ elements from the set $\mathbf{k} - \{i_1, i_2, ..., i_l, j_t\}$. It follows that since $l \leq \mu$, then $|\mathbf{k} - \{i_1, i_2, ..., i_l\}| \geq k - \mu$, and one can find the combination Γ_j. Similarly, it can be shown that the coding sets Ω_i's satisfy the conditions (3.13) and hence are strongly isolating.

Due to the fact that $|\mathbf{k} - \Gamma_i| = \mu$, $i \in \mathbf{p}$, each residual will be decoupled from μ faults, and therefore for the occurrence of concurrent faults in more than μ components, all the residuals will be affected, i.e. for $l > \mu$, we have $\mathbf{p} = \Omega_{i_1 i_2 ... i_l}$. This shows that μ is the maximum number of combinations that satisfy equation (3.12). ■

It can be shown for $\mu = 1$ that the above coding sets result in the generalized residual set [149]. For simplicity, we refer to strongly isolability index as isolability index and drop the term "strongly".

Theorem 3.4 *1. A necessary condition for the SFDIP problem for the linear system* (2.1) *to have a solution for the above coding sets and a given family of fault signatures is that for each $\mu + 1$ combination $L_{i_1}, ..., L_{i_{\mu+1}}$ of L_i's, $[L_{i_1}, ..., L_{i_{\mu+1}}]$ is full rank.*

2. A necessary condition for the SFDIP problem for the nonlinear system (2.34) *to have a solution for the above coding sets and a given family of fault signatures is that for each $\mu + 1$ combination $l_{i_1}(x), ..., l_{i_{\mu+1}}(x)$ of $l_i(x)$'s, the dimension of the distribution $\Delta = span[l_{i_1}(x), ..., l_{i_{\mu+1}}(x)]$ is $v_i = \sum_{j=1}^{\mu+1} k_{i_j}$.*

Proof:

1. If there exists a $\mu + 1$ combination $L_{i_1}, ..., L_{i_{\mu+1}}$ of L_i's such that $[L_{i_1}, ..., L_{i_{\mu+1}}]$ is not full rank, then $\mathcal{L}_{i_{\mu+1}} \subset \sum_{j=1}^{\mu} \mathcal{L}_{i_j}$. It is evident that one of the sets Γ_i is equal to $\mathbf{k} - \{i_1, i_2, ..., i_\mu\}$ and $\mathcal{L}_{i_{\mu+1}} \subset S^*_{\Gamma_i}$, where $S^*_{\Gamma_i} = \inf \mathfrak{S}(A, C, \sum_{j \notin \Gamma_i} \mathcal{L}_j)$ and $\sum_{j \notin \Gamma_i} \mathcal{L}_j = \sum_{j=1}^{\mu} \mathcal{L}_{i_j}$.
2. Similarly, if there exists a $\mu + 1$ combination $l_{i_1}(x), ..., l_{i_{\mu+1}}(x)$ of $l_i(x)$'s such that the dimension of the distribution $\Delta = \text{span}[l_{i_1}(x), ..., l_{i_{\mu+1}}(x)]$ is less than $v_i = \sum_{j=1}^{\mu+1} k_{i_j}$, then there exists $x \in \mathrm{X}$ such that $l_{i_{\mu+1}}(x) \subset \Delta_i = \text{span}\{l_{i_1}(x), ..., l_{i_\mu}(x)\}$. It is evident that one of the sets Γ_i is equal to $\mathbf{k} - \{i_1, i_2, ..., i_\mu\}$ and $\text{span}\{l_{i_{\mu+1}}\} \subset (\Pi^*)^\perp = o.c.a.((\Sigma_*^{\Delta_i})^\perp)$. ■

The above necessary condition provides a test to determine the possible values of the isolability index for a family of fault signatures. For sake of subsequent referencing, this coding scheme is designated as the <u>*coding scheme 1*</u>.

Theorem 3.5 *Let A, C and L_i be arbitrary matrices of dimensions $n \times n$, $q \times n$ and $n \times k_i$, respectively, then the SFDIP problem for the linear system* (2.1) *generically has a solution for the coding sets of Theorem 3.3 if and only if*

1. *For each* $\mu+1$ *combination* $L_{i_1},...,L_{i_{\mu+1}}$ *of* L_i*'s let* $v_i = \sum_{j=1}^{\mu+1} k_{i_j}$ *and* $v = \max_i v_i$*, then*

$$v \leq n \tag{3.17}$$

2. *For each* μ *combination* $L_{i_1},...,L_{i_\mu}$ *of* L_i*'s,*

$$\sum_{j=1}^{\mu} k_{i_j} < q \tag{3.18}$$

Proof: (only if) It readily follows that L_i's should satisfy Theorem 3.4, hence (3.17) is immediate. Moreover, if for a given combination $L_{i_1},...,L_{i_\mu}$ of L_i's, $q < \sum_{j=1}^{\mu} k_{i_j}$, then generically $S^*_{\Gamma_i} = \mathcal{X}$ for $\Gamma_i = \mathbf{k} - \{i_1,...,i_\mu\}$, and therefore condition (3.18) is necessary.

(if) Inequality (3.17) implies that $L_i, i \in \mathbf{k}$ generically satisfies the necessary condition of Theorem 3.4. Also, inequality (3.18) implies that for any μ combination $L_{i_1},...,L_{i_\mu}$ of L_i's, generically $S^*_{\Gamma_i} = \sum_{j=1}^{\mu} \mathcal{L}_{i_j}$. Therefore, from Theorem 3.4 it follows that the SFDIP problem is generically solvable for the coding sets of Theorem 3.3.

■

The above coding scheme needs to generate $C(k, k-\mu)$ residuals for detection and isolation of μ concurrent faults. However, under certain special circumstances one can solve the problem with fewer number of residuals.

Theorem 3.6 *Assume that a given family of fault signatures can be categorized into* m *subsets* $FL_1, FL_2, ..., FL_m$ *such that*

1. *For* $FL_i = \{L_{i_1}, L_{i_2}, ...L_{i_{b_i}}\}$*, then* $[L_{i_1},...,L_{i_{b_i}}]$ *is not full column rank.*
2. *Any* $\mu_i + 1$ *combination of fault signatures in* FL_i *is linearly independent.*

For each subset FL_i*,* $\Gamma_{ij}, j = 1,..,C(b_i, b_i - \mu_i)$ *are defined as* $b_i - \mu_i$ *combinations of* FL_i*, then the coding sets* Ω_i*'s are strongly isolating with index* $\mu = \min_i \mu_i$*.*

Proof: It should be shown that for any two different $1 \leq l, h \leq \mu$ combinations $L_{i_1},...,L_{i_l}$ and $L_{j_1},...,L_{j_h}$, equation (3.12) holds. If both combinations are from one subfamily FL_k, then according to Theorem 3.3, equation (3.12) holds. If combinations are from different subfamilies then by using the fact that for each two fault signatures L_i and L_j from two different subfamilies,

$$\Omega_i \cap \Omega_j = 0 \tag{3.19}$$

it follows that equation (3.12) holds. Let $L_{i_1},...,L_{i_{l_1}}$ and $L_{j_1},...,L_{j_{h_1}}$ be from the same subfamily FL_i and $L_{i_{l_1+1}},...,L_{i_l}$ and $L_{j_{h_1+1}},...,L_{j_h}$ be from a different subfamily FL_j. Then by invoking Theorem 3.3, we have

$$\begin{aligned} \Omega_{i_1,...i_{l_1}} &\neq \Omega_{j_1,...j_{h_1}} \\ \Omega_{i_{l_1+1},...i_l} &\neq \Omega_{j_{h_1+1},...j_h} \end{aligned} \tag{3.20}$$

Also, by using equation (3.19), we have

$$\Omega_{i_1,\dots i_{l_1}} \cap \Omega_{j_{h_1+1},\dots j_h} = 0 \tag{3.21}$$

$$\Omega_{i_{l_1+1},\dots i_l} \cap \Omega_{j_1,\dots j_{h_1}} = 0 \tag{3.22}$$

Therefore, it follows that

$$\Omega_{i_1,\dots i_l} \neq \Omega_{j_1,\dots j_h}$$

The same procedure can be used if $L_{i_1}, \dots, L_{i_l}$ and $L_{j_1}, \dots, L_{j_h}$ can be categorized into more than two subfamilies. ■

The same result can be derived for a nonlinear system with the family of fault signatures $l_i(x)$'s. For the coding sets of Theorem 3.6, the number of residuals are $p = \sum_{i=1}^{m} C(b_i, b_i - \mu_i)$ which may be much smaller than $C(k, \mu)$ when $\mu = \min_i \mu_i$.

Remark 3.2. It should be noted that by utilizing the coding sets of Theorem 3.6, one may detect more than μ concurrent faults in cases where the faults are not all from one of the subfamilies FL_i's. For example, consider the two different $\mu_i + \mu_j$ combinations $L_{i_1}, \dots, L_{i_{\mu_i+\mu_j}}$ and $L_{j_1}, \dots, L_{j_{\mu_i+\mu_j}}$ such that $L_{i_1}, \dots, L_{i_{\mu_i}}$ and $L_{j_1}, \dots, L_{j_{\mu_i}}$ are from the same subfamily FL_i and $L_{i_{\mu_i+1}}, \dots, L_{i_{\mu_i+\mu_j}}$ and $L_{j_{\mu_i+1}}, \dots, L_{j_{\mu_i+\mu_j}}$ are from a different subfamily FL_j, then it follows that

$$\Omega_{i_1,\dots i_{\mu_i+\mu_j}} \neq \Omega_{j_1,\dots j_{\mu_i+\mu_j}}$$

so that one can indeed detect and isolate these two combination of concurrent faults. The maximum number of concurrent faults that one can detect and isolate is $\sum_{i=1}^{m} \mu_i$, where only μ_i fault signatures are selected from the subfamily FL_i.

Theorem 3.7 *1. The SFDIP problem for the linear system* (2.1) *has a solution for the coding sets of Theorem 3.6 and a given family of fault signatures if* $\mathcal{FL}_i \cap L_{k_j} = 0, \forall i, k_j$ *where* $\mathcal{FL}_i = \sum_{j=1}^{b_i} \mathcal{L}_{i_j}$ *and* $L_{k_j} \in FL_k, k \neq i$.
2. The SFDIP problem for the nonlinear system (2.34) *has a solution for the coding sets of Theorem 3.6 and a given family of fault signatures if* $span l_{k_j}(x) \nsubseteq FL_i \; \forall i, k_j$ *where* $FL_i == span\{l_{i_1}(x), \dots, l_{i_{b_i}}(x)\}$ *and* $l_{k_j} \in FL_k, k \neq i$.

Proof:

1. If there exists $L_{k_j} \in FL_k, k \neq i$ such that $\mathcal{FL}_i \cap L_{k_j} \neq 0$ for some $i \in \mathbf{m}$, then there exists an Γ_{kl} such that $k_j \in \Gamma_{kl}$, but since $\mathcal{FL}_i \cap L_{k_j} \neq 0$, then $L_{k_j} \cap S^*_{\Gamma_{kl}} \neq 0$, and therefore the SFDIP problem does not have a solution for Γ_{kl}.
2. The proof is similar to part 1 and is omitted. ■

It should be mentioned that in the above coding set, the residuals for each subfamily should be decoupled from all faults in all other subfamilies.

It will be shown in the next section that actuator fault signatures in a network of unmanned vehicles satisfy the above decoupling property. The next example illustrates some details regarding the above coding scheme. For subsequent referencing, the coding scheme in Theorem 3.6 is designated as the *coding scheme 2.*

Example 3.1. Consider the following family of fault signatures

$$FL = \begin{bmatrix} 1 & -0.5 & -0.5 & 0 & 0 & 0 & 0 & 0 \\ 0 & 0 & 0 & 0 & 1 & -0.5 & -0.5 & 0 \\ 0 & 0.5 & -0.5 & 0 & 0 & 0 & 0 & 0 \\ 0 & 0 & 0 & 0 & 0 & 0.5 & -0.5 & 0 \\ 0 & 0 & -0.5 & 0.5 & 0 & 0 & 0 & 0 \\ 0 & 0 & 0 & 0 & 0 & 0 & 0.5 & -0.5 \end{bmatrix} \tag{3.23}$$

One can categorize the above family into two sets FL_1 and FL_2 such that

$$FL_1 = \begin{bmatrix} 1 & -0.5 & -0.5 & 0 \\ 0 & 0 & 0 & 0 \\ 0 & 0.5 & -0.5 & 0 \\ 0 & 0 & 0 & 0 \\ 0 & 0 & -0.5 & 0.5 \\ 0 & 0 & 0 & 0 \end{bmatrix} = \begin{bmatrix} L_1 & L_2 & L_3 & L_4 \end{bmatrix} \tag{3.24}$$

$$FL_2 = \begin{bmatrix} 0 & 0 & 0 & 0 \\ 1 & -0.5 & -0.5 & 0 \\ 0 & 0 & 0 & 0 \\ 0 & 0.5 & -0.5 & 0 \\ 0 & 0 & 0 & 0 \\ 0 & 0 & 0.5 & -0.5 \end{bmatrix} = \begin{bmatrix} L_5 & L_6 & L_7 & L_8 \end{bmatrix} \tag{3.25}$$

where $\mu_1 = \mu_2 = 2$ and $b_1 = b_2 = 4$ as defined in Theorem 3.6. The sets Γ_{ij} are selected as $\Gamma_{11} = \{1,2\}$, $\Gamma_{12} = \{1,3\}$, $\Gamma_{13} = \{1,4\}$, $\Gamma_{14} = \{2,3\}$, $\Gamma_{15} = \{2,4\}$, $\Gamma_{16} = \{3,4\}$, $\Gamma_{21} = \{5,6\}$, $\Gamma_{22} = \{5,7\}$, $\Gamma_{23} = \{5,8\}$, $\Gamma_{24} = \{6,7\}$, $\Gamma_{25} = \{6,8\}$, and $\Gamma_{26} = \{7,8\}$. Utilizing the coding scheme of Theorem 3.6, one needs to generate only 12 residuals, whereas by utilizing the coding scheme of Theorem 3.3, $C(8,2) = 28$ residuals need to be generated. Moreover, by assuming that there are concurrent faults in L_1, L_2, L_5, L_6, then all the residuals except r_{16} and r_{26} will be affected by this fault, with r_{ij} denoting the residual generated by Γ_{ij}. Therefore, one can detect and isolate the occurrence of 4 concurrent faults in these fault signatures.

3.3 Actuator Fault Detection and Isolation in a Network of Unmanned Vehicles

In the following discussions, the proposed structured residual set is utilized to solve the actuator fault detection and isolation problem in a network of unmanned vehicles. We will investigate the development of and comparison among two architectures for FDI of actuator faults in the network of unmanned vehicles. Namely, we are interested in centralized and semi-decentralized configurations as described and defined in Sections 3.1.1 and 3.1.3, respectively.

3.3.1 Centralized Architecture

As pointed out in Section 3.1.1, system (3.6) is overactuated and the column rank of $\bar{B}^O$ in system (3.6) is $a \times (N-1)$. Hence, the first step is to determine the isolability index of actuator fault signatures in the centralized architecture. It can be verified that the family of $a \times N$ fault signatures $\bar{L}^O_{kj}, k \in \mathbf{N},\ j \in \mathbf{a}$ satisfies the necessary condition of Theorem 3.4 with $\mu = N-2$ isolability index. If the SFDIP problem has a solution for the *coding scheme 1*, then one needs to generate $C(a \times N, N-2)$ residuals for detecting and isolating $N-2$ multiple faults. However, since matrix B is full rank and in view of the structure of the matrix $\bar{B}^O$, one can categorize the family of fault signatures into the following a subfamilies $FL_i, i \in \mathbf{a}$ such that $FL_i = \{\bar{L}^O_{ki}, k \in \mathbf{N}\}$ satisfies the condition of Theorem 3.6 with $b_i = N$ and $\mu_i = N-2$. In other words, each subfamily FL_i contains the fault signatures of the i-th actuator of all the vehicles. If the SFDIP problem has a solution for the *coding scheme 2*, then one needs to generate $a \times C(N, N-2) = a \times N \times (N-1)/2$ residuals for detecting and isolating $N-2$ multiple faults, which is much smaller than $C(a \times N, N-2)$. Moreover, according to Remark 3.2, one is able to detect multiple faults in all actuators of each vehicle since fault signatures of each vehicle are in different subfamilies.

Theorem 3.8 presented below provides the necessary and sufficient conditions for generating the residual signals for the centralized architecture by using the *coding scheme 2*. The following lemma is needed and will be used in the proof of this theorem.

Lemma 3.2 *Consider the following diagonal dynamical system*

$$\begin{aligned} \dot{x}(t) &= A^M x(t) + B^M u(t) + L^M m(t) \\ y(t) &= C^M x(t) \end{aligned} \tag{3.26}$$

where $L = [L_1, \cdots, L_a]$ *and* $M > 1$ $(A^M = I_M \otimes A,\ B^M = I_M \otimes B,\ L^M = I_M \otimes L)$. *Let* $L_{ij}, i \in \boldsymbol{M} = \{1, ..., M\}, j \in \boldsymbol{a}$ *denote the* $(i-1)\times a+j$*-th column of* L^M*, then the family of fault signatures* L_{ij}*'s in* (3.26) *is strongly detectable if and only if the family of fault signatures* L_i*'s is strongly detectable in the following system*

$$\begin{aligned} \dot{x}(t) &= Ax(t) + Bu(t) + \sum_{k=1}^{a} L_k m_k(t) \\ y(t) &= Cx(t) \end{aligned} \tag{3.27}$$

Proof: The fault signatures L_{ij}'s are strongly detectable if and only if there exists (C^M, A^M) unobservability subspace $\mathcal{S}_{ij}^{M^*} = \inf \underline{\mathcal{S}}^M(\bar{\mathcal{L}}_{ij}^M)$ such that $\mathcal{S}_{ij}^{M^*} \cap \mathcal{L}_{ij} = 0$, where $\bar{L}_{ij}^M, i \in \mathbf{M}, j \in \mathbf{k}$ are obtained by setting the column L_{ij} of L^M to zero. By using the Algorithms 2.1 and 2.3, it can be easily shown that $\mathcal{S}_{ij}^{M^*} = \text{diag}\{\mathcal{S}^*(\mathcal{L}), \mathcal{S}^*(\mathcal{L}), \cdots, \mathcal{S}^*(\bar{\mathcal{L}}_j), \cdots, \mathcal{S}^*(\mathcal{L})\}$ where $\bar{L}_j, j \in \mathbf{k}$ are obtained by setting the j-th column of L to zero and $\mathcal{S}^*(\mathcal{L}) = \inf \underline{\mathcal{S}}(\mathcal{L})$ is (C, A) unobservability subspace. Hence, $\mathcal{S}_{ij}^{M^*} \cap \mathcal{L}_{ij} = 0$ if and only if $\mathcal{S}^*(\bar{\mathcal{L}}_j) \cap \mathcal{L}_j = 0$, implying that L_j's should be strongly detectable. ■

We are now in the position to introduce our main theorem in this subsection.

Theorem 3.8 *The centralized SFDIP problem has a solution for the* coding scheme 2 *if and only if the actuator fault signatures* L_k*'s are strongly detectable for the system* (3.27).

Proof: According to Theorem 3.6, for each subfamily FL_i, the sets Γ_{ij} are defined as the 2-combination of N fault signatures corresponding to the i-th actuator of each vehicle. Therefore, each residual r_{ij} should only be affected by the i-th actuator fault signatures of the two vehicles and should be decoupled from all other actuator fault signatures. Since the observable state x^O is considered as the relative state of the vehicles with respect to the state of vehicle 1, we consider generating the residual signals r_{ij}'s such that the residual signal r_{ij} is affected by the i-th actuator of vehicle 1 and vehicle j and is decoupled from all other actuator fault signatures in the entire system. In generating such residual signals, system (3.6) can be rewritten as follows

$$\begin{aligned} \dot{x}^O(t) &= A^{N-1}x^O(t) + B^{N-1}u^O(t) + L^{N-1}m^O(t) \\ \bar{z}(t) &= C^{N-1}x^O(t) \end{aligned} \tag{3.28}$$

where

$$u^O = \begin{bmatrix} u_1(t) - u_2(t) \\ \vdots \\ u_1(t) - u_N(t) \end{bmatrix}, m^O(t) = \begin{bmatrix} m_1(t) - m_2(t) \\ \vdots \\ m_1(t) - m_N(t) \end{bmatrix}, m_i(t) = \begin{bmatrix} m_{i1}(t) \\ \vdots \\ m_{ia}(t) \end{bmatrix}.$$

Using the result of the Lemma 3.2, it follows that the residual signals r_{ij}'s can be generated if and only if L_j's are strongly detectable in system (3.27). The same approach can be considered for generating the residual signals that are only affected by the i-th actuator fault signatures of vehicles $k, j \neq 1$ by rewriting the observable subsystem in terms of the relative states with respect to the vehicle k. ■

In the next section, it is shown that by considering only "local" communication links among neighboring vehicles, the vehicles with more than one neighbor can detect and isolate not only their own faults but also faults of their neighboring vehicles.

3.3.2 Semi-decentralized Architecture

As pointed out in Section 3.1.3, the family of actuator fault signatures in the nodal model (3.10) is dependent and system (3.10) is also overactuated. It can be verified that the family of $a \times (|\mathrm{N}_i| + 1)$ fault signatures $\bar{L}^O_{kj}, k = 1, ..., |\mathrm{N}_i| + 1,\ j \in \mathbf{a}$ in system (3.10) satisfies the necessary condition of Theorem 3.4 with the isolability index of $\mu = |\mathrm{N}_i| - 1$.

If the SFDIP problem has a solution for the *coding scheme 1*, then one needs to generate $C(a \times (|\mathrm{N}_i| + 1), |\mathrm{N}_i| - 1)$ residuals for detecting and isolating $|\mathrm{N}_i| - 1$ multiple faults. However, since matrix B is full rank, and by considering the structure of the matrix $\bar{B}^O_{\mathrm{N}_i}$, one can categorize the family of fault signatures into the following a subfamilies $FL_j, j \in \mathbf{a}$ such that $FL_j = \{\bar{L}^O_{kj}, k = 1, ..., |\mathrm{N}_i| + 1\}$ satisfies the conditions of Theorem 3.6 with $b_j = |\mathrm{N}_i| + 1$ and $\mu_j = |\mathrm{N}_i| - 1$. If the SFDIP problem has a solution for the *coding scheme 2*, then one needs to generate $a \times C(|\mathrm{N}_i| + 1, |\mathrm{N}_i| - 1) = a \times (|\mathrm{N}_i| + 1) \times |\mathrm{N}_i|/2$ residuals for detecting and isolating $|\mathrm{N}_i| - 1$ multiple faults, which is much less than $C(a \times (|\mathrm{N}_i| + 1), |\mathrm{N}_i| - 1)$. Moreover, according to Remark 3.2, vehicles with more than one neighbor are able to detect multiple faults in all their actuators since fault signatures of each vehicle are in different subfamilies.

Remark 3.3. It should be noted that for the vehicles with only one neighbor ($|\mathrm{N}_i| = 1$), the isolability index is $\mu = 0$, which implies that one cannot isolate any fault in the vehicle and its neighbor. However, since the matrix B is full rank, one can still categorize the fault signatures into a subfamily $FL_j, j \in \mathbf{a}$ such that $FL_j = \{\bar{L}^O_{1j}, -\bar{L}^O_{1j}\}$. In this case according to the coding scheme 2, one needs to design a residual generators such that each one is affected by only one subfamily and is decoupled from others. By utilizing this coding scheme, one can isolate faults among actuators but cannot isolate between the vehicle and its only neighbor.

The result of this subsection is now summarized in the following theorem.

Theorem 3.9 *The semi-decentralized SFDIP problem has a solution for the* coding scheme 2, *if the fault signatures L_k's are strongly detectable for the system* (3.27).

Proof: Proof is similar to that of Theorem 3.8 and is omitted.

According to Theorems 3.8 and 3.9, the solvability conditions for the FDI problem in the centralized and semi-decentralized architectures are the same. Therefore, for vehicles with more than one neighbor, the same isolability performance will be achieved by using either scheme. However, in the semi-decentralized scheme the FDI algorithms are distributed among the vehicles and the computational load of each vehicle is much less than that of the centralized FDI unit.

In order to compare the total number of residuals that are necessary in the centralized and the semi-decentralized architectures, let us assume that each vehicle has at most two neighbors. This assumption is based on the fact that a vehicle having two neighbors can detect and isolate its own faults. With this assumption in mind, the total number of residuals in the semi-decentralized architecture is found to be at most $3aN$, which for $N > 7$ is less than the total number of residuals needed in the centralized architecture which is $a \times N \times (N-1)/2$. This shows that for a relatively "large" network of unmanned vehicles, that is for $N > 7$, the total computational load of the semi-decentralized architecture is also much less than the centralized approach.

It should be noted that in the semi-decentralized architecture, the residuals are distributed among the vehicles. Moreover, by utilizing the semi-decentralized architecture, one can also reduce the practical limitations that are often associated with the centralized architecture vis-*á*-vis computational bottlenecks and scarce communication bandwidth resources. Consequently, the semi-decentralized FDI algorithm is more scalable and flexible when compared to the centralized architecture. Note that in the semi-decentralized architecture the communication overhead per each vehicle is higher than the centralized approach. However, one should also keep in mind that the efforts required (due to the presence of obstacles, limited line of sight, etc.) to maintain the communication links between each vehicle and the central FDI unit in the centralized approach are much higher than those needed to keep the communication links among only the neighboring vehicles in the semi-decentralized architecture.

3.3.3 Simulation Results For Formation Flight of Satellites

In this section, the proposed fault detection and isolation strategy is applied to an application area of significant strategic interest, namely the satellite precision formation flight problem [159, 40]. For simulation studies, the relative motion of four satellites are configured with respect to a reference virtual satellite that is following a desired orbit and that can be approximated by linearizing the Keplerian orbital mechanics about this reference trajectory. These equations are known as the Hill-Clohessy-Wiltshire equations [161]. The equations of motion for each satellite about a circular reference orbit are governed by the following model

$$\begin{aligned} \ddot{x} &= 3n^2x + 2n\dot{y} + u_x \\ \ddot{y} &= -2n\dot{x} + u_y \\ \ddot{z} &= -n^2z + u_z \end{aligned} \tag{3.29}$$

where x points in the radial direction, y points along the track, and z points out of the plane. The orbital rate is given by n. The vector $[u_x^\top,\ u_y^\top,\ u_z^\top]^\top$ represents external input accelerations due to the applied actuators thrust. It is assumed that each satellite can measure its relative position with respect to its neighboring satellites where in this example we have assumed $N_1 = \{2\}$, $N_2 = \{1,3\}$, $N_3 = \{4\}$ and $N_4 = \{1\}$. The satellites are tasked to form a regular square. It follows from equation (3.29) that the xy-dynamics are decoupled from the z-dynamics. Here, we only consider actuator faults in the xy-dynamics.

A multiple fault scenario is considered for simulations where four faults are injected in both actuators of satellites 2 and 4. Specially, (a) the x-direction and the y-direction actuators of satellite 4 are injected with a lock-in-place fault resulting in the actuators being locked at the values of $u_{41} = 0.1$ and $u_{42} = -0.1$ at $t = 10$ and $t = 15$ seconds, respectively, (b) an 80% loss of effectiveness fault is injected at the x-direction actuator of satellite 2 at $t = 20$ seconds, and (c) the y-direction actuator of satellite 2 is locked at the value of $u_{21} = 0.5$ at $t = 20$ seconds.

One can easily verify that the actuator fault signatures of each satellite are strongly detectable when each satellite has exact position measurements (that is $z_i = Cx_i$). Therefore, according to the Theorem 3.8 the centralized FDI architecture has a solution for the coding scheme 2 and the isolability index of the entire family of signatures is 2. Actuator fault signatures of all the satellites can be categorized into two sets $FL_1 = [\bar{L}^O_{11}, \bar{L}^O_{21}, \bar{L}^O_{31}, \bar{L}^O_{41}]$ and $FL_2 = [\bar{L}^O_{12}, \bar{L}^O_{22}, \bar{L}^O_{32}, \bar{L}^O_{42}]$, where $b_1 = b_2 = 4$, $\mu_1 = \mu_2 = 2$, and $\bar{L}^O_{ki}$ represents the fault signature of the i-th actuator of the k-th satellite in the observable part of the entire network. A total of 12 residuals $r_i, i = 1, ..., 12$ are needed for the coding scheme 2. The family of the coding sets for the above residual set and the family of fault signatures are as follows $\Omega^c_{11} = \{1,2,3\}$, $\Omega^c_{21} = \{1,4,5\}$, $\Omega^c_{31} = \{2,5,6\}$, $\Omega^c_{41} = \{3,4,6\}$, $\Omega^c_{12} = \{7,8,9\}$, $\Omega^c_{22} = \{7,10,11\}$, $\Omega^c_{32} = \{8,11,12\}$ and $\Omega^c_{42} = \{9,10,12\}$, where Ω^c_{ij} corresponds to the fault signal m_{ij}.

In the semi-decentralized architecture, since satellites 1, 3 and 4 have only one neighbor, according to Remark 3.3 they can isolate faults among actuators but cannot isolate between themselves and their only one neighbor. By utilizing the coding scheme 2, two residuals r_{i1} and r_{i2} are designed for the i-th satellite ($i = 1, 3, 4$) where they are affected by faults in the x-direction actuator and the y-direction actuator of the satellite and its neighbor, respectively. However, since $|N_2| = 2$, satellite 2 not only can detect and isolate multiple faults in its own actuator but also it can detect and isolate faults in satellites 1 and 3. The actuator fault signatures of satellites 1, 2, and 3 can be categorized into two sets $FL_{21} = [\bar{L}_{11}^O, \bar{L}_{21}^O, \bar{L}_{31}^O]$ and $FL_{22} = [\bar{L}_{12}^O, \bar{L}_{22}^O, \bar{L}_{32}^O]$, where $b_1 = b_2 = 3$, $\mu_1 = \mu_2 = 1$, and $\bar{L}_{ki}^O$ represents the fault signature of the i-th actuator of the k-th satellite in the observable part of the nodal model of satellite 2. Based on the coding scheme 2, six residual signals $r_{2j}, j = 1, ..., 6$ are needed for satellite 2. Therefore, in the semi-decentralized architecture, a total of 12 residuals should be generated. Residual evaluation function for each residual is selected as an instantaneous value of the residual, i.e. $J_{r_i} = r_i$. A uniformly random noise (5%) is added to the relative distance measurements (representing the sensor noise). By considering the worst case analysis of the residuals corresponding to the healthy operation of the satellites that are subject to measurement noise, a threshold value of ± 0.1 is selected for all the residual signals for fault detection and isolation logic evaluation and analysis. The family of the coding sets for satellite 2 is as follows: $\Omega_{11} = \{21, 22\}$, $\Omega_{21} = \{21, 23\}$, $\Omega_{31} = \{22, 23\}$, $\Omega_{12} = \{24, 25\}$, $\Omega_{22} = \{24, 26\}$, and $\Omega_{32} = \{25, 26\}$, where Ω_{ij} corresponds to the fault signal m_{ij}.

Figures 3.1 and 3.2 show the residuals corresponding to the considered fault scenario for the centralized architecture. As shown in this figure, faults in the x and y direction actuators of satellite 4 can be detected and isolated by using Ω_{41}^c and $\Omega_{41}^c \cup \Omega_{42}^c$ at $t = 10.5$ and $t = 16.4$ seconds, respectively. The occurrence of faults in the x and y direction actuators of satellite 2 can be detected and isolated by using $\Omega_{21}^c \cup \Omega_{41}^c \cup \Omega_{42}^c$, and $\Omega_{21}^c \cup \Omega_{22}^c \cup \Omega_{41}^c \cup \Omega_{42}^c$ at $t = 20.5$ and $t = 25.6$ seconds, respectively. Figures 3.3 and 3.4 show the residuals that are generated by using our detection filters associated with the considered fault scenario for the semi-decentralized architecture. As shown in these figures, satellite 2 can detect and isolate the fault in its x-direction actuator at $t = 20.5$ seconds by using the coding set Ω_{21}, and the fault in its y-direction actuator at $t = 25.8$ seconds by using the coding set $\Omega_{21} \cup \Omega_{22}$, but satellites 1, 3, and 4 can only detect the occurrence of faults in both actuators of the satellites and cannot isolate among the satellites. It should be noted that all the residual signals in both centralized and semi-decentralized architectures are generated according to the geometric approach introduced in Chapter 2.

Based on the above results one can conclude that for satellites with more than one neighbor (satellite 2 in the above formation example) the detection and isolation performance of the centralized and the semi-decentralized architectures are similar (multiple faults in all their actuators can be detected and isolated). However, in the semi-decentralized architecture the FDI algo-

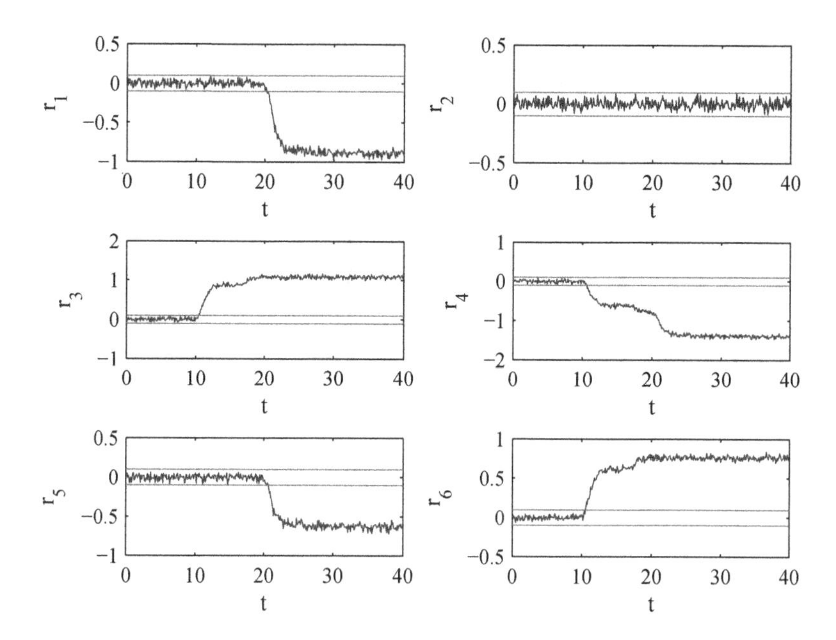

Fig. 3.1 The residuals corresponding to multiple fault scenario $r_1, ... r_6$ (centralized architecture).

rithm is distributed among the satellites. This results in a solution that is more robust and fault-tolerant to complete failure of a central FDI unit that is generally placed in one of the satellites (a single point of failure).

3.4 Fault Detection and Isolation of an F-18 HARV (High Angle of Attack Research Vehicle)

To demonstrate the effectiveness of the proposed fault detection and isolation strategy for overactuated systems, we consider actuator faults in an F-18 High Angle of Attack Research Vehicle (HARV) [122, 129]. For simplicity, a linear longitudinal motion is considered. The state vector is defined as $x^\top = [v, \alpha, q, \theta]$, where v represents a change in the forward speed of the aircraft, α represents a change in the angle of attack, q represents a change in the pitch rate and θ represents a change in the pitch angle. There are six independent input channels, namely, (1) δ_{TH}: throttle change (2) δ_{TV}: thrust vectoring change (3) δ_{STAB}: change in stabilator, (4) δ_{ASYM}: change in symmetrical aileron, (5) δ_{LEF}: change in leading edge flap deflection, and (6) δ_{TEF}: change in trailing edge flap deflection.

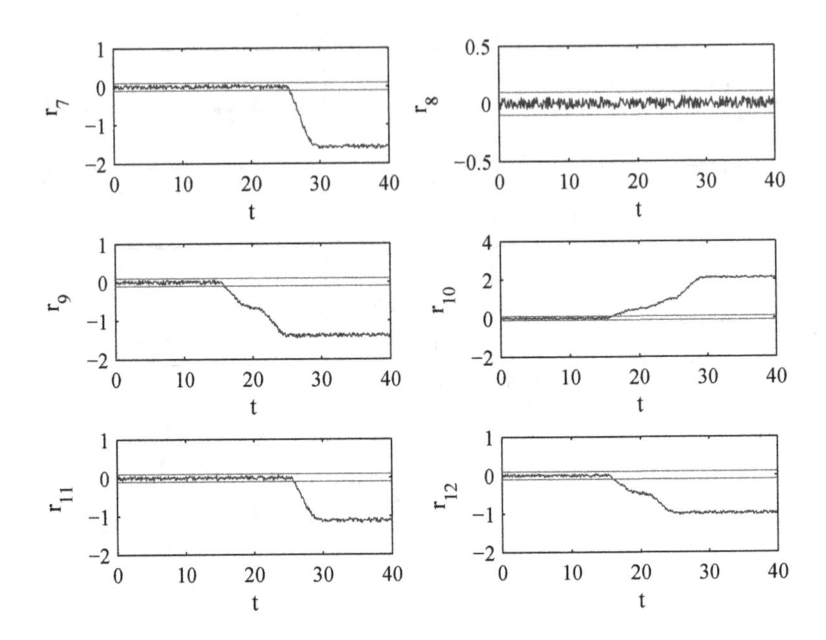

Fig. 3.2 The residuals corresponding to multiple fault scenario $r_7, ..., r_{12}$ (centralized architecture).

The aircraft is assumed to be flying at the Mach number of 0.24 at an altitude of 15000 ft which corresponds to a steady forward speed of 239 ft/s. The equilibrium flight path angle is 0° and the angle of attack is 25°. The state equation corresponding to this flight condition is given by

$$\dot{x} = Ax + Bu, \quad y = x \tag{3.30}$$

where

$$A = \begin{bmatrix} -0.075 & -24.05 & 0 & -32.16 \\ -0.0009 & -0.196 & 0.9896 & 0 \\ -0.0002 & -0.1454 & -0.1677 & 0 \\ 0 & 0 & 1.0 & 0 \end{bmatrix}$$

$$B = \begin{bmatrix} -1.15 & 0 & -2.482 & 0.0393 & -2.466 & 4.32 \\ -0.01 & -0.005 & -0.0136 & 0 & -0.018 & -0.008 \\ -0.335 & -0.035 & -0.408 & -0.0006 & -0.042 & 0.0135 \\ 0 & 0 & 0 & 0 & 0 & 0 \end{bmatrix}$$

and

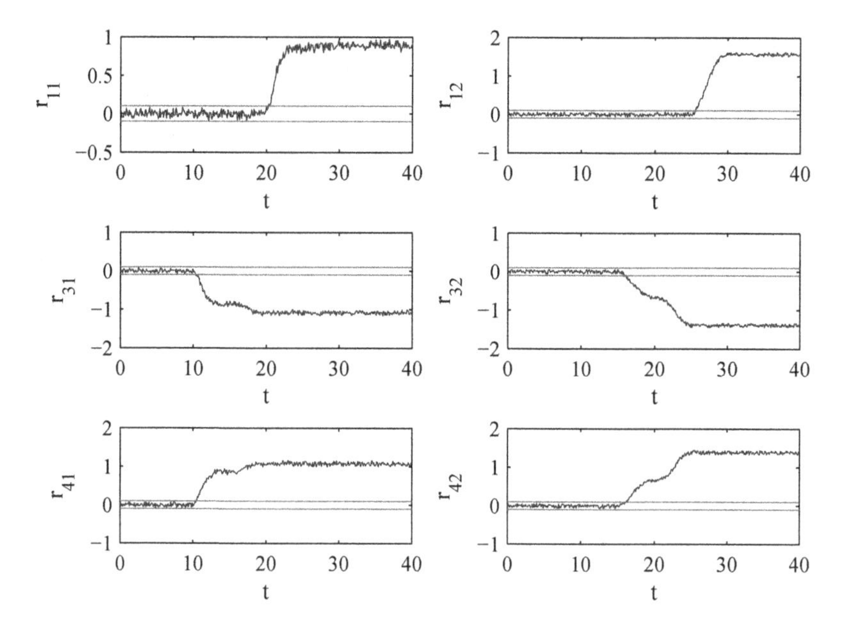

Fig. 3.3 The residuals corresponding to multiple fault scenario (semi-decentralized architecture).

$$u = \left[\delta_{TH}\ \delta_{TV}\ \delta_{STAB}\ \delta_{ASYM}\ \delta_{LEF}\ \delta_{TEF}\right]^T$$

According to Theorem 3.4, the possible value of the isolability index for this system is 2. One can check that the observability subspaces of Theorem 3.3 also exist for $\mu = 2$, and therefore in addition to detecting faults in each input channel one can also detect concurrent faults in two input channels. Based on the coding scheme 1, a total of 15 ($C(6,2)$) residuals may be generated such that each is decoupled from 2 input channels. It can be verified that the fault signatures cannot be categorized according to the coding scheme 2.

By taking into account the maximum value or the worst case scenario of residuals corresponding to the normal mode of the system subject to measurement noise and by allowing some safety margins, a threshold value of 0.1 was considered for all the residuals. The coding sets for the above fault signatures are as follows

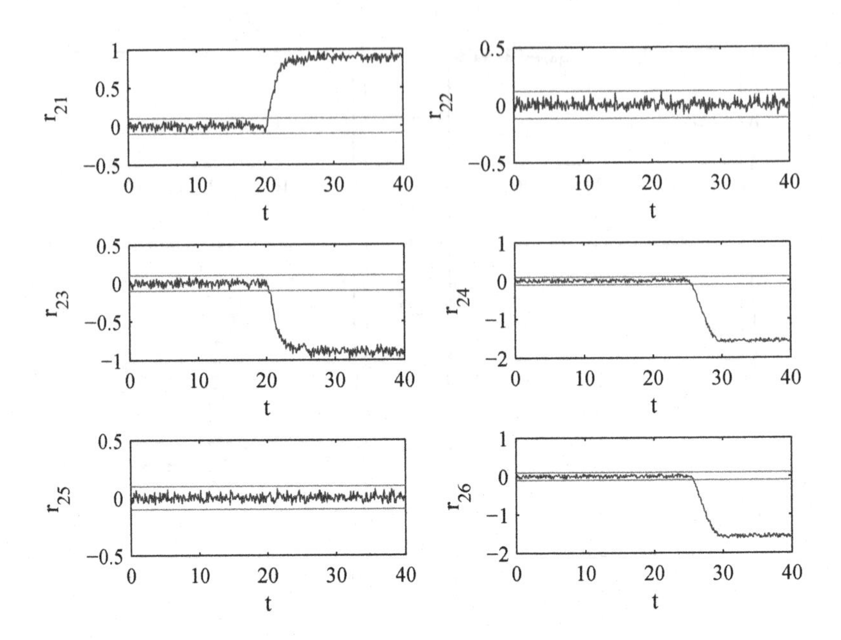

Fig. 3.4 The residuals corresponding to multiple fault scenario (semi-decentralized architecture).

$$\begin{aligned}
\Omega_1 &= \{6, 7, 8, 9, 10, 11, 12, 13, 14, 15\} \\
\Omega_2 &= \{2, 3, 4, 5, 10, 11, 12, 13, 14, 15\} \\
\Omega_3 &= \{1, 3, 4, 5, 7, 8, 9, 13, 14, 15\} \\
\Omega_4 &= \{1, 2, 4, 5, 6, 8, 9, 11, 12, 15\} \\
\Omega_5 &= \{1, 2, 3, 5, 6, 7, 9, 10, 12, 14\} \\
\Omega_6 &= \{1, 2, 3, 4, 6, 7, 8, 10, 11, 13\}
\end{aligned}$$

where Ω_i corresponds to the fault signatures of the $i-th$ actuator.

First, we consider that there is a hard-over fault in the input channel $u_5 = \delta_{LEF}$ at $t = 50$ sec. Figure 3.5 depicts the residual outputs corresponding to this fault scenario. According to this figure one can detect and isolate the fault by using the coding set Ω_5. Next, we consider a 60% loss of effectiveness fault in the input channel $u_6 = \delta_{TEF}$ at $t = 50$ sec. Figure 3.6 shows the residual outputs corresponding to this fault scenario. According to this figure one can clearly detect this fault by using Ω_6. Finally, we consider concurrent hard over faults in the two input channels, namely $u_2 = \delta_{TH}$ at $t = 50$ and $u_5 = \delta_{LEF}$ at $t = 70$ sec. Figures 3.7 shows the residual outputs corresponding to this scenario. According to these graphs, one can detect and isolate the concurrent faults from the coding set $\Omega_2 \cup \Omega_5$.

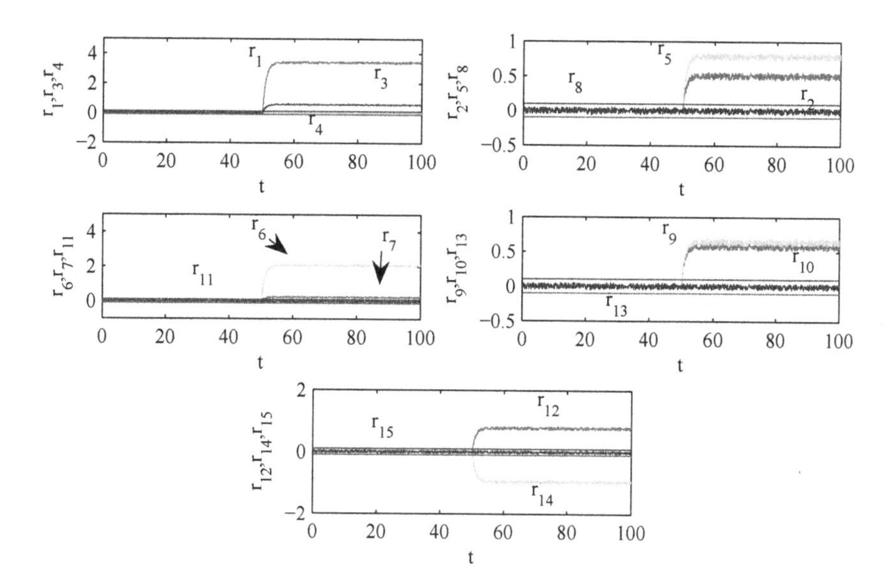

Fig. 3.5 Residual outputs corresponding to the hard-over fault in the leading edge flap deflection.

3.5 Fault Detection and Isolation of Redundant Reaction Wheels of a Satellite

Let $\mathcal{F}_o$ denote the inertial frame with origin $O \in \mathcal{R}^3$ and $\mathcal{F}_b$ a moving body frame whose origin O_i is at the mass center of the spacecraft. The attitude kinematics may be described by the quaternion $q = (\hat{q}^\top, q_4)$. The attitude and angular velocity of the spacecraft with respect to the inertial frame $\mathcal{F}_o$ can be described by the following quaternion equation [103, 128]

$$\begin{aligned} \frac{{}^o d\hat{q}}{dt} &= \frac{q_4\omega - \omega \times \hat{q}}{2} \\ \frac{q_4}{dt} &= -\frac{\omega \cdot \hat{q}}{2} \end{aligned} \tag{3.31}$$

and the Euler equation

$$\begin{aligned} &\frac{{}^o d(I\omega + h)}{dt} = \tau_d \\ &I\frac{{}^b d\omega}{dt} + \omega \times (I\omega + h) = -\dot{h} + \tau_d \end{aligned} \tag{3.32}$$

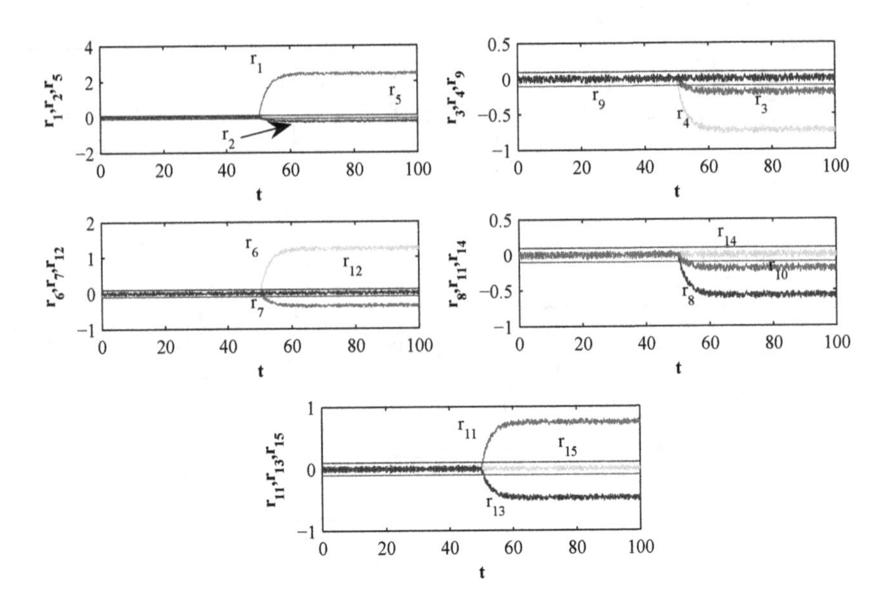

Fig. 3.6 Residual outputs corresponding to 60% loss of effectiveness in the trailing edge flap deflection.

where $q^\top = (\hat{q}^\top, q_4)$ denotes the unit quaternion, I is the tensor of inertia in the body frame $\mathcal{F}_b$, h is the angular momentum of the wheels, and τ_d is the external disturbance torque that is associated with the spacecraft.

Let $h_i, (i = 1, 2, 3, 4)$ denotes the angular momentum of the ith reaction wheel and $h_w = [h_1, h_2, h_3, h_4]^\top$, then the relationship between h and h_w is given by

$$h = Ah_w \tag{3.33}$$

where $A \in \mathcal{R}^{3\times4}$ is the configuration matrix of the reaction wheels. Four wheels in a tetrahedron configuration are employed where A is given by [103]

$$A = \begin{bmatrix} \sqrt{\frac{1}{3}} & \sqrt{\frac{1}{3}} & -\sqrt{\frac{1}{3}} & -\sqrt{\frac{1}{3}} \\ \sqrt{\frac{2}{3}} & -\sqrt{\frac{2}{3}} & 0 & 0 \\ 0 & 0 & -\sqrt{\frac{2}{3}} & \sqrt{\frac{2}{3}} \end{bmatrix} \tag{3.34}$$

By substituting equation (3.33) into equation (3.32), one yields

$$I\frac{{}^b d\omega}{dt} + \omega \times (I\omega + Ah_w) = -Au + \tau_d \tag{3.35}$$

where the control input is $u = \dot{h}_w$. We further assume that the following output measurements are available:

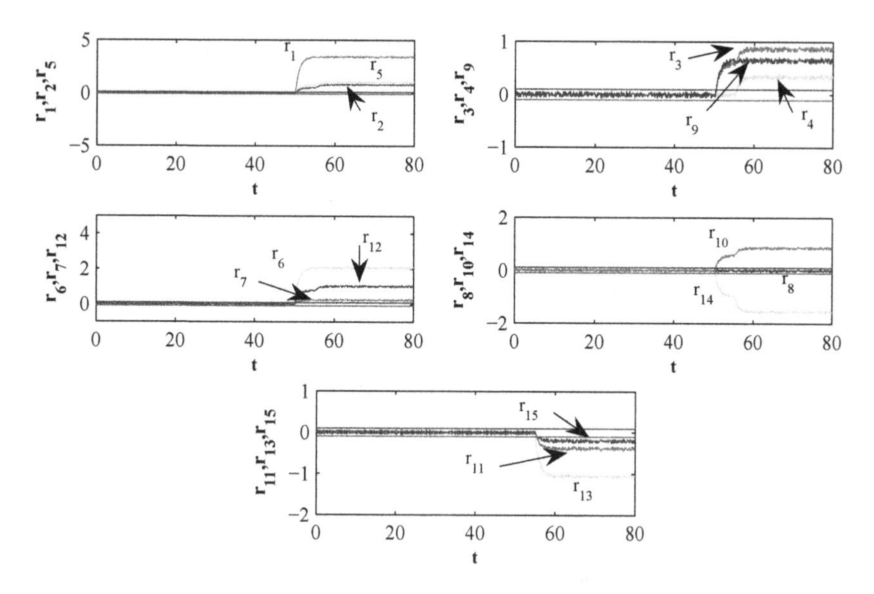

Fig. 3.7 Residual outputs corresponding to concurrent hard-over faults in the throttle and the trailing edge flap deflections.

$$y = \begin{bmatrix} \omega \\ h_w \end{bmatrix} \tag{3.36}$$

The main problem that we are addressing here is how one can detect and isolate faults in the redundant reaction wheels (the minimum number of wheels required for the spacecraft is three that are associated with the pitch, yaw, and the roll angles). We can rewrite the equation (3.35) as

$$\begin{aligned} \dot{\omega} &= f(\omega) + \sum_{i=1}^{4} g_i u_i + \tau_d \\ y &= \omega \end{aligned} \tag{3.37}$$

where

$$f(\omega) = -I^{-1}(\omega \times (I\omega + Ah_w)),$$

$$g_1 = I^{-1} \begin{bmatrix} \sqrt{\frac{1}{3}} \\ \sqrt{\frac{2}{3}} \\ 0 \end{bmatrix}, g_2 = I^{-1} \begin{bmatrix} \sqrt{\frac{1}{3}} \\ -\sqrt{\frac{2}{3}} \\ 0 \end{bmatrix}$$

$$g_3 = I^{-1} \begin{bmatrix} -\sqrt{\frac{1}{3}} \\ 0 \\ -\sqrt{\frac{2}{3}} \end{bmatrix}, g_4 = I^{-1} \begin{bmatrix} -\sqrt{\frac{1}{3}} \\ 0 \\ \sqrt{\frac{2}{3}} \end{bmatrix}$$

According to Theorem 3.4, one can check that the isolability index for the above system is 2. Thus one should try to generate $C(4,2) = 6$ residual signals $r_i(t), i = 1, ..., 6$. The sets Γ_i's for having the isolability index of 2 can be found from Theorem 3.3 (coding scheme 1) as $\Gamma_1 = \{3,4\}$, $\Gamma_2 = \{2,4\}$, $\Gamma_3 = \{2,3\}$, $\Gamma_4 = \{1,4\}$, $\Gamma_5 = \{1,3\}$, $\Gamma_6 = \{1,2\}$ and the corresponding coding sets are $\Omega_1 = \{4,5,6\}$, $\Omega_2 = \{2,3,6\}$, $\Omega_3 = \{1,3,5\}$, and $\Omega_4 = \{1,2,3\}$.

Next, the observability codistributions of Theorem 2.6 for the above coding sets are obtained by using the algorithm that is proposed in [154] as follows

$$\begin{aligned}
\Pi^*_{\Gamma_1} &= \text{span}\{d\omega_3\} \\
\Pi^*_{\Gamma_2} &= \text{span}\{d(-0.8165\omega_1 + 0.4619\omega_2 + 0.3464\omega_3\} \\
\Pi^*_{\Gamma_3} &= \text{span}\{d(0.8165\omega_1 - 0.4619\omega_2 + 0.3464\omega_3\} \\
\Pi^*_{\Gamma_4} &= \text{span}\{d(-0.8165\omega_1 - 0.4619\omega_2 + 0.3464\omega_3\} \\
\Pi^*_{\Gamma_5} &= \text{span}\{d(0.8165\omega_1 + 0.4619\omega_2 + 0.3464\omega_3\} \\
\Pi^*_{\Gamma_6} &= \text{span}\{d(\omega_2)\}
\end{aligned}$$

where it is assumed that the inertia matrix is $I = \text{diag}\{5,4,3\}$. It is easy to verify that the necessary conditions (2.47) of Theorem 2.6 are satisfied and the SFDIP problem has a solution for the above system with an isolability index of 2. According to the above observability codistributions, the following set of new coordinates can be found

$$\begin{aligned}
z_1 &= \omega_3 \\
z_2 &= -0.8165\omega_1 + 0.4619\omega_2 + 0.3464\omega_3 \\
z_3 &= 0.8165\omega_1 - 0.4619\omega_2 + 0.3464\omega_3 \\
z_4 &= -0.8165\omega_1 - 0.4619\omega_2 + 0.3464\omega_3 \\
z_5 &= 0.8165\omega_1 + 0.4619\omega_2 + 0.3464\omega_3 \\
z_6 &= \omega_2
\end{aligned}$$

The state space representation of the satellite's attitude dynamics may now be expressed in the new coordinates as follows

$$
\begin{aligned}
\dot{z}_1 =& \frac{1}{3}(\omega_1\omega_2 + \omega_2 h_{w_1} - \omega_1 h_{w_2} + \sqrt{\frac{2}{3}}(u_4 - u_3)) \\
\dot{z}_2 =& -0.8165(\frac{1}{5}(\omega_2\omega_3 + \omega_3 h_{w_2} - \omega_2 h_{w_3})) \\
&+ 0.4619(\frac{1}{4}(-2\omega_1\omega_3 - \omega_3 h_{w_1} + \omega_1 h_{w_3})) \\
&+ 0.3464(\frac{1}{3}(\omega_1\omega_2 + \omega_2 h_{w_1} - \omega_1 h_{w_2})) + 0.1886(u_4 - u_2) \\
\dot{z}_3 =& 0.8165(\frac{1}{5}(\omega_2\omega_3 + \omega_3 h_{w_2} - \omega_2 h_{w_3})) \\
&- 0.4619(\frac{1}{4}(-2\omega_1\omega_3 - \omega_3 h_{w_1} + \omega_1 h_{w_3})) \\
&+ 0.3464(\frac{1}{3}(\omega_1\omega_2 + \omega_2 h_{w_1} - \omega_1 h_{w_2})) + 0.1886(u_2 - u_3) \\
\dot{z}_4 =& -0.8165(\frac{1}{5}(\omega_2\omega_3 + \omega_3 h_{w_2} - \omega_2 h_{w_3})) \\
&- 0.4619(\frac{1}{4}(-2\omega_1\omega_3 - \omega_3 h_{w_1} + \omega_1 h_{w_3})) \\
&+ 0.3464(\frac{1}{3}(\omega_1\omega_2 + \omega_2 h_{w_1} - \omega_1 h_{w_2})) + 0.1886(u_4 - u_1) \\
\dot{z}_5 =& 0.8165(\frac{1}{5}(\omega_2\omega_3 + \omega_3 h_{w_2} - \omega_2 h_{w_3})) \\
&- 0.4619(\frac{1}{4}(-2\omega_1\omega_3 - \omega_3 h_{w_1} + \omega_1 h_{w_3})) \\
&+ 0.3464(\frac{1}{3}(\omega_1\omega_2 + \omega_2 h_{w_1} - \omega_1 h_{w_2})) + 0.1886(u_1 - u_3) \\
\dot{z}_6 =& \frac{1}{4}(-2\omega_2\omega_3 - \omega_3 h_{w_1} + \omega_1 h_{w_3} + \sqrt{\frac{2}{3}}(u_1 - u_2))
\end{aligned}
$$

where

$$
\begin{bmatrix} h_{w1} \\ h_{w2} \\ h_{w3} \end{bmatrix} = A h_w \tag{3.38}
$$

The next step is to design observers for generating the estimated states $\hat{z}_1$, $\hat{z}_2$, $\hat{z}_3$, $\hat{z}_4$, $\hat{z}_5$, and $\hat{z}_6$, and subsequently define the residuals as $r_i = z_i - \hat{z}_i$, $i \in \{1, ..., 6\}$.

The simulation results of the proposed nonlinear FDI scheme when applied to the attitude dynamics of the satellite are presented below. The circular dawn-dusk sun synchronous orbit with a 6 pm ascending node and 650 km altitude has been considered for simulations. The satellite is subjected to four disturbance torques, namely: the gravity gradient, the solar radiation, the Earth's magnetic field, and the aerodynamic torque. Table 3.1 shows the characteristics of the disturbance torques for the selected orbit. The Earth observation mission is considered for the satellite with pointing accuracy of 0.5 degree in all the three axes. According to the mission requirements and disturbances, a PD controller was designed for the satellite. One of the common faults in the reaction wheel is the loss of effectiveness due to the bus voltage drop or friction [4, 173, 8, 110, 180] which can be modeled as

$$u_{true} = ku_c, \qquad 0 \leq k < 1 \tag{3.39}$$

where u_{true} and u_c denote the actual output and the controller command of the reaction wheel, respectively.

Table 3.1 Orbital Disturbances

Disturbances	Magnitude	Constant/cyclic
Gravity gradient	$3.006e-8$ N.m	Constant
Solar radiation	$5.468e-6$ N.m	Constant
The Earth's magnetic field	$4.568e-5$ N.m	Cyclic
Aerodynamic torque	$7.008e-7$ N.m	Constant

Figure 3.8 shows the residuals corresponding to the normal (healthy) operation of the satellite. According to this figure, the residuals are cyclic due to disturbances. First, we consider a 20% loss of effectiveness in the reaction wheel #3. Figure 3.9 depicts the residuals corresponding to this fault. As shown in this figure, the residuals are still cyclic, but the magnitude of the residuals is increased due to the presence of the fault. This cyclic feature of the residuals will make the fault decision making process rather difficult and not straightforward. To remedy this problem, the L_2 norm of the residuals in each orbit is considered for deciding on the fault detection and isolation tasks. Figure 3.10 shows the L_2 norm of the residuals corresponding to the healthy mode where the norms are reset when the satellite passes the perigee of its orbit. According to Figure 3.10 different threshold values are chosen for each residual as shown in Table 3.2. Figure 3.11 depicts the norm of the residuals corresponding to the 20% loss of effectiveness fault in the reaction wheel #3. It can easily be observed that one detects and isolates the fault in this reaction wheel by using the coding set Ω_3. Finally, a concurrent fault scenario is also considered for the reaction wheels #1 and #3. Figures 3.12 and 3.13 show the residuals and their norms corresponding to the faulty scenario. As shown in Figure 3.13, all residuals except r_2 exceed their thresholds and based on the coding set $\Omega_1 \cup \Omega_3$ one can detect and isolate both faults in the reaction wheels.

Table 3.2 Residual threshold values

Residuals	Threshold Value
r_1	0.01
r_2	0.0005
r_3	0.002
r_4	0.002
r_5	0.012
r_6	0.005

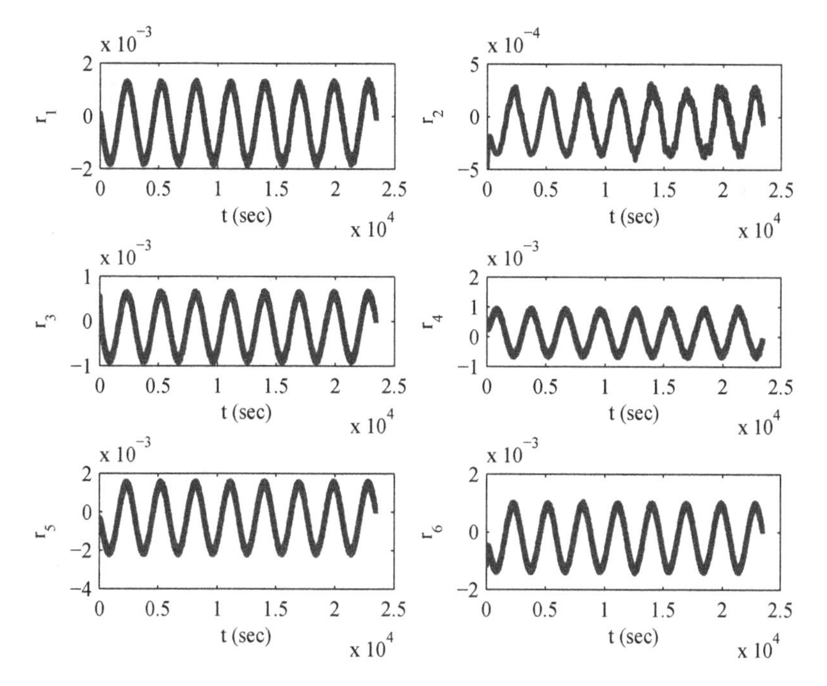

Fig. 3.8 Residuals corresponding to the normal mode (healthy operation).

3.6 Conclusions

In this chapter, new structured residual sets are designed and developed for both linear and nonlinear systems with dependent fault signatures. The notion of an isolability index is formally introduced and necessary and sufficient conditions for the coding sets to achieve a specific isolability index is obtained. The proposed residual set is applied to three case studies, namely, the actuators FDI problem in an F-18 HARV aircraft, the actuator FDI problem in a satellite with redundant reaction wheel and the FDI problem in a network of unmanned vehicles. Three FDI architectures, namely, centralized, decentralized and semi-decentralized are considered for a network of unmanned

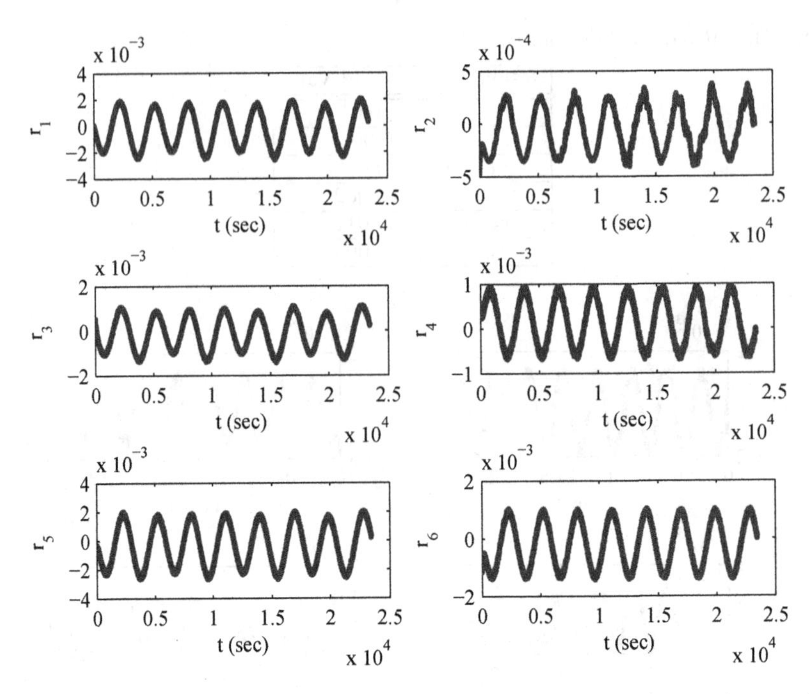

Fig. 3.9 Residuals corresponding to the 20% loss of effectiveness fault in the reaction wheel #3 (©IEEE 2007), reprinted with permission.

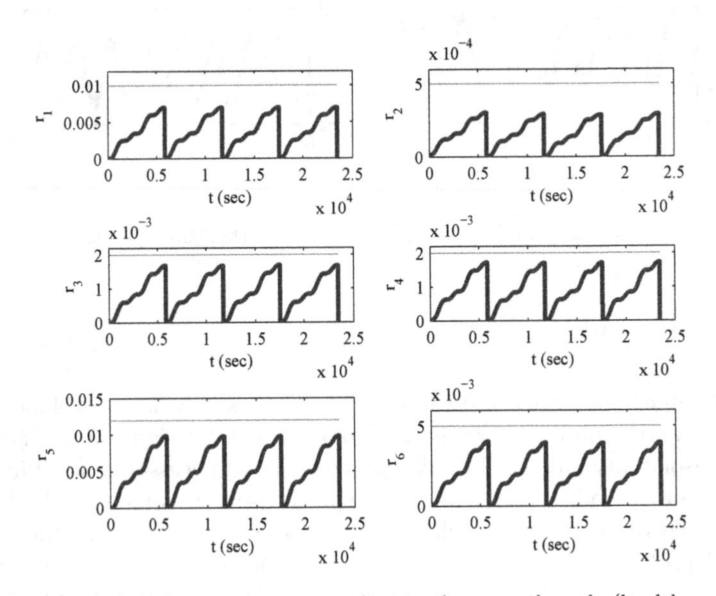

Fig. 3.10 Residual L_2 norm corresponding to the normal mode (healthy operation) (©IEEE 2007), reprinted with permission.

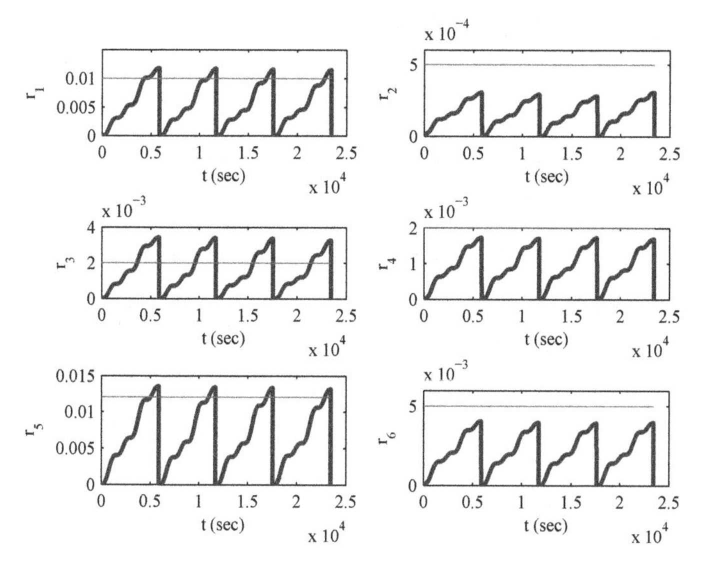

Fig. 3.11 Residual L_2 norm corresponding to the 20% loss of effectiveness fault in the reaction wheel # 3 (©IEEE 2007), reprinted with permission.

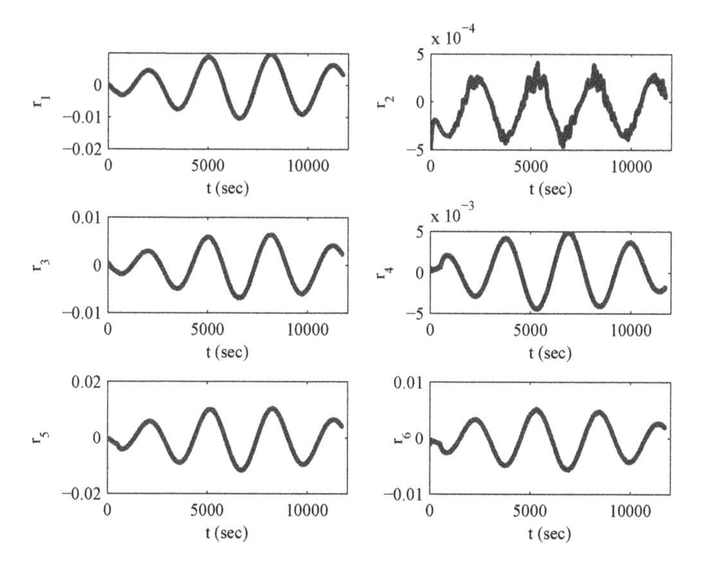

Fig. 3.12 Residuals corresponding to the 40% loss of effectiveness fault in the reaction wheel #3 and the 50% loss of effectiveness fault in the reaction wheel #1 (©IEEE 2007), reprinted with permission.

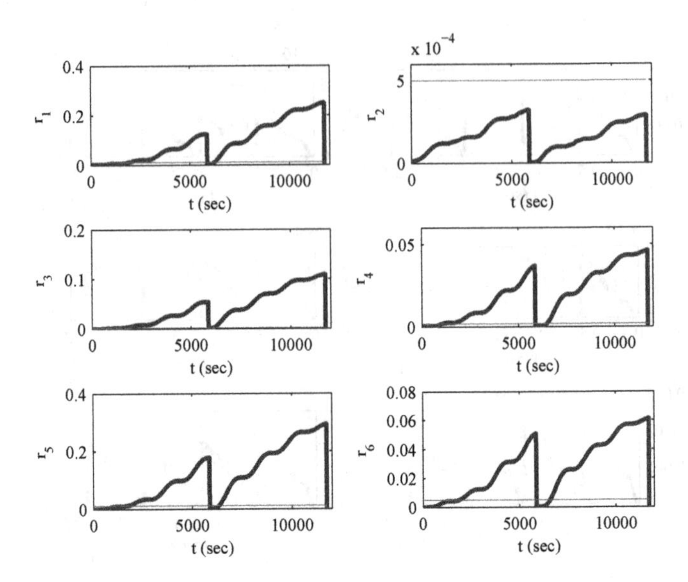

Fig. 3.13 Residuals L_2 norm corresponding to the 40% loss of effectiveness fault in the reaction wheel #3 and the 50% loss of effectiveness fault in the reaction wheel #1 (©IEEE 2007), reprinted with permission.

vehicles. Simulation results are also presented to illustrate and demonstrate the effectiveness of the proposed approach.

Chapter 4
A Robust FDI Scheme with a Disturbance Decoupling Property

In this chapter, the actuator fault detection and isolation problem for a network of unmanned vehicles subject to large input disturbances is considered. One of the main challenges in the design of FDI algorithms is to distinguish the effects of disturbances from faults and develop a robust FDI scheme without compromising the detection of incipient faults in the vehicles. In unmanned vehicles such as UAV's, this problem is more challenging due to the small size feature and higher sensitivity of these vehicles to disturbances such as wind gust. In this chapter, a robust FDI scheme is designed by developing a hybrid fault detection and isolation strategy for both linear and nonlinear systems that are subject to large environmental disturbances. The work presented in this chapter has partially appeared in [136, 138, 135].

This chapter is organized as follows. We begin with a brief literature review. In Section 4.2, we formulate the FDI problem for both linear and nonlinear systems subject to external disturbances. In Section 4.2.1, a systematic approach for generating a novel set of residuals in a hybrid FDI scheme is presented. In Section 4.2.2, the residual evaluation functions that are required for the hybrid FDI scheme are discussed. A discrete-event system (DES) fault diagnoser is presented in Section 4.2.3. In Sections 4.3 and 4.4, the proposed hybrid FDI algorithm is applied to the actuators fault detection and isolation problem in a network of unmanned vehicles (linear systems) and an Almost-Lighter-Than-Air-Vehicle (ALTAV) (nonlinear systems), respectively.

4.1 Introduction

One of the important issues in the model-based FDI is robustness against uncertainties, unmodeled dynamics, and modeling errors. Since a perfectly accurate and complete mathematical model of a physical system is never available and parameters of the system may vary with time in an uncertain manner, there is always a mismatch between the actual process and its mathematical model even when there is no process faults. These discrepancies cause fundamental complications and difficulties in the FDI applications. Therefore, considering the effects of modeling uncertainties is among the most crucial issues in the model-based FDI problem. Various robust model-based FDI techniques have been developed in the literature. These include Unknown Input Observer (UIO) approach [186, 65, 66, 165], eigenstructure assignment [147, 30, 148], optimal parity relation for robust FDI [112, 193, 57], frequency domain design and H_∞ optimization FDI approach [183, 67, 98]. A comprehensive treatment of the robust FDI problem is presented in [31].

In UIO robust FDI approach, disturbances and model uncertainties are modeled as an additive disturbance in the state space representation of the system. Model uncertainties include coupling and interconnecting terms in large scale systems, nonlinear terms in system dynamics, terms arising from time-varying system dynamics, linearization and model reduction errors, and parameters variations. In this approach, an observer is designed whose estimation error converges to zero regardless of an unknown input (disturbance) in the system. The remaining design degrees of freedom are used for fault detection and isolation purposes. In the eigenstructure assignment robust FDI, a direct approach for designing disturbance decoupled residuals is developed where the residual signal is decoupled from the disturbances while the state estimation error may not be decoupled from it. The existing conditions for an eigenstructure assignment problem can be relaxed when compared with those that are required for the UIO.

In the optimal parity relation approach, two objective functions for the design of parity relations are defined. The optimization objectives are the minimization of the modeling uncertainty effects and the maximization of the fault effects. This leads to a multi-objective optimization problem which is solved by forming a mixed objective function optimization problem. In this approach the uncertainty is considered as bounded parameter variations and a set of possibilities for system parameters within their bounds is considered to describe multiple models of the system. In an H_∞ optimization approach one tries to keep the sensitivity of the residual signal to unknown inputs (disturbances) less than a specific bound while increase the sensitivity of the residual signal to the fault over the frequency range of the fault. Different methods can be used to achieve the above criteria such as coprime factorization, LMI method, and H_∞ filtering based on algebraic Riccati equation. The main difference between the UIO and the eigenstructure approaches with parity relation and H_∞ approaches is that in the first two approaches one seeks to decouple the effect of unknown inputs from the residual signals. However, in the second pair of robust approaches, effects of unknown inputs on residual signals are dynamically attenuated. Therefore, solvability conditions of the second pair of approaches are generally more relaxed when compared to the first two approaches.

In all the above approaches, the modeling uncertainties and errors are considered as unknown inputs and one seeks to decouple or attenuate the effects of them on the residuals. An alternative approach [208, 207] is to consider structured model uncertainty in the system and design a robust fault detection filter for the system with such structured uncertainty. However, in this approach prior choice of a nominal model can lead to suboptimal solutions in all situations in which a nominal plant model is not easily identifiable or is not available. To remedy this problem, recently [27, 97] have developed robust fault detection filters for systems with polytopic model uncertainty based on an H_∞ optimization approach.

In another research direction, the discrete-event based fault diagnosis approaches [158, 83, 84] attempt to represent the nominal and the faulty-system behaviors in the form of event trajectories, and to design fault diagnosers to estimate the current status of the system. This abstraction may result in a loss of information that can be critical to the task of fault isolation. Recently, hybrid fault diagnosis approaches have been investigated for complex dynamical systems whose behavior is modeled as a hybrid system. In [123], fault diagnosis of a continuous-time system with embedded supervisory controllers subject to abrupt, partial and full failure of components is investigated as a model selection problem. Reference [139] presents an online model-based diagnosis methodology for parametric faults in hybrid systems, which is based on tracking hybrid behavior (consisting of both continuous-time and discrete-event behaviors). In [49] a discrete-event system diagnosis approach is presented for abrupt parametric faults in continuous-time systems based on qualitative abstraction of the system behavior.

In this chapter, a novel hybrid FDI algorithm for both linear and nonlinear systems that are subject to large environmental disturbances is developed. Many modern systems such as aircraft and balloons are expected to operate in harsh environments where large disturbances (wind gust) may be present. One of the main challenges in designing a robust FDI algorithm for these systems is to determine how to distinguish between large environmental disturbances and faults or failures.

Towards the above end, a hybrid architecture for a robust FDI is introduced that is composed of a bank of continuous-time residual generators and a DES fault diagnoser. First a set of residual signals is generated based on the coding set that is introduced in Chapter 3 for a family of fault signatures with a given isolability index. Two threshold levels are assigned to the residual signals. It is further assumed that the input disturbances can be categorized into two families, namely, the tolerable disturbance inputs and the large and unexpected disturbances. A first level of threshold is selected such that the tolerable disturbance inputs do not generate any false alarms by using the residual signals. Next, a complementary set of residual signals is generated by considering the effects of the disturbances on the first set of generated residual signals. A DES fault diagnoser is then designed to invoke an appropriate combination of the residual signals and their sequential features to not only detect and isolate faults and guarantee no false alarms subject to large external disturbance signals, but also to detect and identify the occurrence of large external disturbances. It should be emphasized that the proposed FDI approach performs simultaneous robust fault detection and isolation as well as large and unexpected disturbances detection without imposing any limitations on the total number of faults that can be detected and isolated. In contrast, in the previous FDI algorithms developed in [164, 154, 19], robustness against disturbance inputs is achieved by limiting the number of possible faults that can be present in the system. The proposed hybrid FDI algorithm is subsequently applied to the actuators fault detection and isolation problem in a network of unmanned vehicles as well as an Almost-Lighter-Than-Air-Vehicle (ALTAV).

4.2 Hybrid FDI Approach

Consider the following linear system

$$\begin{aligned} \dot{x}(t) &= Ax(t) + Bu(t) + \sum_{i=1}^{k} L_i m_i(t) + \sum_{j=1}^{P} P_j \omega_j(t) \\ y(t) &= Cx(t) + v(t) \end{aligned} \tag{4.1}$$

and the nonlinear system

$$\begin{aligned} \dot{x} &= f(x) + g(x)u + \sum_{i=1}^{k} l_i(x) m_i + \sum_{j=1}^{P} p_j(x) \omega_j \\ y &= h(x) + v \end{aligned} \tag{4.2}$$

where $\omega_j \in \mathbb{R}^{P_j}$ denotes the disturbance input, v represents the measurement noise and L_i and $l_i(x)$ represent fault signatures. It is assumed that $\omega_j, v \in \mathfrak{L}^p[0, \infty]$ for some $1 \leq p \leq \infty$ where $\mathfrak{L}^p[0, \infty]$ denotes the space of $\mathfrak{L}^p$ norm bounded signals, i.e. $||\omega||_p < \infty$.

Assumption 4.1 *The disturbance inputs are categorized into two types, namely tolerable disturbance signals* $\mathfrak{D}_1 = \{\omega \in \mathfrak{L}^p[0, \infty] \mid ||\omega||_p < \delta_1\}$ *and large and unexpected disturbance signals* $\mathfrak{D}_2 = \{\omega \in \mathfrak{L}^p[0, \infty] \mid \delta_1 \ll ||\omega||_p < \delta_2\}$ *where* $\delta_1 \ll \delta_2$.

Assumption 4.2 *The faults and the large disturbance inputs do not occur simultaneously and there exists a sufficient time separating the occurrence of a fault and the disturbance.*

The objective of this chapter is to design a Hybrid Fault Diagnoser (HFD) for detecting and isolating each fault m_i while guaranteeing that the diagnoser remains robust with respect to both types of disturbances. In other words, no false alarms should be generated due to the disturbance signals.

The hybrid fault diagnoser is composed of two modules, namely, a low-level bank of residual generators and a high level DES diagnoser. The bank of continuous-time residual generators produces first a set of residual signals based on the geometric FDI approach. It then compares, by using an evaluation function, each residual signal to its corresponding threshold value, from which a set of residual logic units is generated. Two levels of thresholds are needed for certain residual signals (this will be discussed in more details subsequently). The DES diagnoser module is a finite-state automaton that takes the residual logic units as inputs and estimates the current state of the system. For designing such a DES diagnoser, the combined plant and the bank of residual generators is modeled as a finite state Moore automaton (G). The general architecture of the proposed hybrid fault diagnoser is shown in Figure 4.1.

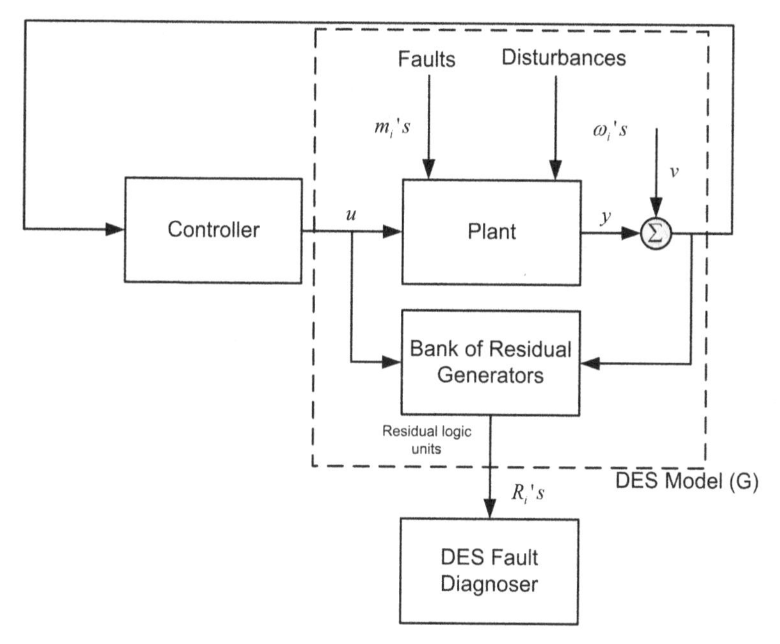

Fig. 4.1 The general architecture of the proposed hybrid diagnoser.

Remark 4.1. One possible approach to design a robust FDI algorithm for system (4.1) ((4.2)) is to generate a set of residual signals [120] ([154]) that each residual is affected by only one fault and is decoupled from all other faults and all the disturbances. If such a set of residual signals exist, then one can robustly detect and isolate faults despite the presence of disturbances. Under these circumstances, there will be no need to have a hybrid structure for the fault diagnoser. However, such a set of residual signals cannot be always generated due to the presence of large number of faults in the system.

In the following sections, the procedure for designing a hybrid fault diagnoser that is composed of a bank of residual generators and a DES diagnoser is described in detail.

4.2.1 Bank of Continuous-Time Residual Generators

In this section, a systematic approach is proposed to design a set of residual generators that provides the necessary information required by the DES diagnoser. Towards this end, two sets of residual signals are developed. The first set is generated according to the coding set that is introduced in Chapter 3 for a family of fault signatures with a given isolability index. The hybrid fault diagnoser (HFD) developed below is guaranteed to remain robust with respect to both tolerable disturbance inputs $\omega_i \in \mathfrak{D}_1$ and measurement noise v by selecting appropriate threshold values associated with the first set of residual signals. To ensure that the HFD is also robust to large disturbance inputs ($\omega_i \in \mathfrak{D}_2$), a second set of complementary residuals is generated so that the DES fault diagnoser, by utilizing the entire set of residual signals, will robustly detect and isolate a fault.

In the following, we assume that the isolability index of L_i's ($l_i(x)$'s) is $\mu \leq k$ (which includes strong isolability). Therefore, the SFDIP problem has a solution for the coding scheme Ω_i's with isolability index of μ, and a set of residuals $r_i, i \in \Xi_1 = \{1, ..., \xi\}$ can be generated ($\xi = C(k, \mu)$ for the coding scheme 1). We denote $\mathfrak{R}_1 = \{r_i, i \in \Xi_1\}$. Let Λ_i denote the set of disturbance signatures P_j's ($p_j(x)$'s) that affects the residual r_i. In other words, for the linear system (4.1),

$$\Lambda_i = \{j \in \mathbf{P} | \mathcal{P}_j \cap \mathcal{S}^*_{\Gamma_i} = 0\}, \quad i \in \Xi_1 \tag{4.3}$$

and for the nonlinear system (4.2)

$$\Lambda_i = \{j \in \mathbf{P} | \operatorname{span}\{p_j(x)\} \not\subseteq (\Pi^*_{\Gamma_i})^{\perp}\}, \quad i \in \Xi_1 \tag{4.4}$$

where

$$\mathcal{S}^*_{\Gamma_i} = \inf \mathfrak{S}(A, C, \sum_{j \notin \Gamma_i} \mathcal{L}_j) \tag{4.5}$$

and

$$\Pi^*_{\Gamma_i} = \text{o.c.a.}((\Sigma_*^{\mathcal{L}_{\Gamma_i}})^{\perp}). \tag{4.6}$$

Assume that one can generate a set of complementary residuals $\mathfrak{R}_2 = \{r_{\xi+i}, i \in \Xi_1\}$ such that $r_{\xi+i}$ is decoupled from the disturbance inputs specified by Λ_i but is affected by all the faults $m_l, l \in \Gamma_i$ and possibly other fault modes. Based on the geometric framework that was presented in Chapter 3 for both linear and nonlinear systems, the next two theorems provide the necessary and sufficient conditions for existence of the residuals $r_i \in \mathfrak{R}_2$ for linear and nonlinear systems, respectively.

Theorem 4.1 *The residual signals $r_{i+\xi} \in \mathfrak{R}_2$ can be generated if and only if*

$$\mathcal{S}^*_{\xi+i} \cap \mathcal{L}_j = 0, \quad j \in \Gamma_i, i \in \Xi_1 \tag{4.7}$$

where $\mathcal{S}^*_{\xi+i} = \inf \underline{\mathcal{S}}(\sum_{j \in \Lambda_i} \mathcal{P}_j)$.

Proof: The proof follows from Theorem 2.3 and is omitted. ■

Theorem 4.2 *The residual signals $r_i \in \mathfrak{R}_1 \cup \mathfrak{R}_2$ can be generated if and only if there exist observability codistributions $\Pi^*_{\Gamma_i} = o.c.a.((\Sigma_*^{\mathcal{L}_{\Gamma_i}})^\perp), i = 1, ..., 2\xi$ where*

$$\mathcal{L}_{\Gamma_i} = span\{l_j(x), j \notin \Gamma_i\}, \quad i \in \Xi_1 \tag{4.8}$$
$$\mathcal{L}_{\Gamma_{\xi+i}} = span\{p_j(x), j \in \Lambda_i\}, \quad i \in \Xi_1 \tag{4.9}$$

such that

$$span\{l_j\} \not\subseteq [\Pi^*_{\Gamma_i}]^\perp, \quad i \in \Xi_1 \tag{4.10}$$
$$span\{l_j\} \not\subseteq [\Pi^*_{\Gamma_{\xi+i}}]^\perp, \quad i \in \Xi_1 \tag{4.11}$$

for all $j \in \Gamma_i$.

Proof: The proof follows from Theorem 2.6 and is omitted. ■

For each disturbance input $\omega_j \in \mathbf{P}$ and the fault mode $m_i, i \in \mathbf{k}$, the coding sets Ω^p_j and Ω^f_i are defined respectively as follows (for linear systems)

$$\Omega^p_j = \{i \in \{1, ..., \xi\} | \mathcal{S}^*_{\Gamma_i} \cap \mathcal{P}_j = 0\} \cup \{i \in \{\xi+1, ..., 2\xi\} | \mathcal{S}^*_i \cap \mathcal{P}_j = 0\} \tag{4.12}$$

and $\Omega^f_i = \Omega_i \cup \Upsilon^f_i$ where $\Upsilon^f_i = \{j \in \{\xi+1, ..., 2\xi\} \mid \mathcal{S}^*_j \cap \mathcal{L}_i = 0\}$. In other words, the sets Ω^p_j and Ω^f_i are the index set of those residuals $r_k \in \mathfrak{R}_1 \cup \mathfrak{R}_2$ that are affected by p_j and m_i, respectively. For each fault $m_i, i \in \mathbf{k}$, the set Υ^f_i represents the complementary residuals $r_i \in \mathfrak{R}_2$ that are affected by m_i.

Assumption 4.3 *For disturbance inputs ω_j such that $j \notin \bigcup_{i=1}^{\xi} \Lambda_i$, it is assumed that $\Omega^p_j = \emptyset$.*

The disturbances which satisfy Assumption 4.3 have no effect on the residuals, and therefore the hybrid diagnoser does not need to be robust to them. In other words, the generated set of residuals are already decoupled from these disturbances and no further invoking of the DES diagnoser is required. The following two lemmas will be used subsequently to design the proposed DES fault diagnoser.

Lemma 4.1 *a) The coding sets Ω_i^p and Ω_j^f are distinct, i.e. $\Omega_i^p \neq \Omega_j^f, i \in \mathbf{P}, j \in \mathbf{k}$, and b) The coding sets Ω_i^f and $\Omega_j^f, i \neq j$ are distinct, i.e. $\Omega_i^f \neq \Omega_j^f, i, j \in \mathbf{k}$, $i \neq j$.*

Proof: a) First we consider the disturbances ω_j, $j \in \mathbf{P}$ such that $j \in \Lambda_i$ for some $i \in \Xi_1$ (ω_j affects at least one of the residuals $r_i \in \mathfrak{R}_1$). Since the residual $r_{\xi+i}$ is decoupled from ω_j and is affected by all the faults $m_i, i \in \Gamma_i$, we have $\xi + i \notin \Omega_j^p$ and $\xi + i \in \Omega_l^f$, $l \in \Gamma_i$. Hence, we have $\Omega_j^p \neq \Omega_l^f$, $l \in \Gamma_i$. Moreover, for all m_l, $l \in \mathbf{k}$ such that $l \notin \Gamma_i$, we have $i \in \Omega_j^p$ and $i \notin \Omega_l^f$; hence $\Omega_j^p \neq \Omega_l^f$, $l \notin \Gamma_i$. Therefore, $\Omega_j^p \neq \Omega_i^f, i \in \mathbf{k}$. Next, we consider the disturbances ω_j, $j \in \mathbf{P}$ such that $j \notin \cup_{i=1}^{\xi} \Lambda_i$, i.e. disturbance inputs that do not affect any of the residuals $r_i, i \in \Xi_1$. According to Assumption 4.3, we have $\Omega_j^p = 0$. However, for any $j \in \mathbf{k}$, there exists at least one residual signal $r_l \in \mathfrak{R}_1$ ($l \in \Xi_1$) such that $l \in \Omega_j^f$; hence $\Omega_j^p \neq \Omega_i^f, i \in \mathbf{k}$.
b) Given the procedure in Chapter 3 for generating the residual signals r_i, $i \in \Xi_1$, we conclude that $\Omega_i \neq \Omega_j$, $i, j \in \mathbf{k}$, $i \neq j$, and we have $\Omega_i^f = \Omega_i \cup \Upsilon_i^f, i \in \mathbf{k}$. Since $\Omega_i \cap \{\xi + 1, ..., 2\xi\} = 0$, $i \in \mathbf{k}$ and $\Upsilon_i^f \subset \{\xi + 1, ..., 2\xi\}$, it follows that $\Omega_i^f \neq \Omega_j^f$, $i, j \in \mathbf{k}$, $i \neq j$. ■

The above lemma ensures that the occurrence of faults and disturbances can be uniquely distinguished.

Lemma 4.2 *Consider the sets Υ_i^f as defined above, then $\Omega_j^p \neq \Upsilon_i^f$, $i \in \mathbf{k}, j \in \mathbf{P}$.*

Proof: For the disturbance inputs $\omega_j \in \mathbf{P}$ such that $j \in \Lambda_i$ for some $i \in \mathbf{k}$, the proof follows along the same lines as that in the proof of part a) of Lemma 4.1. According to Assumption 4.3, for the disturbance inputs that do not affect any of the residual signals $r_i \in \mathfrak{R}_1$ we have $\Omega_j^p = \emptyset$. Since $i + k \in \Upsilon_i^f$, then $\Omega_j^p \neq \Upsilon_i^f$. ■

As will be demonstrated in Section 4.2.3, the above lemma guarantees that one can distinguish between the occurrence of low severity faults and the disturbances.

Remark 4.2. For a system where $\Lambda_i = \Lambda, i \in \mathbf{k}$, in other words when the set of disturbances that affects all the residuals $r_i \in \mathfrak{R}_1$ are the same, only a single extra residual $r_{\xi+1}$ is sufficient for designing the hybrid FDI scheme. According to Theorem 4.2, the residuals can be generated for the nonlinear system if and only if there exist observability codistributions $\Pi_{\Gamma_i}^* = o.c.a.((\sum_*^{\mathcal{L}_i})^\perp), i = 1, ..., \xi + 1$ where

$$\mathcal{L}_{\Gamma_i} = \text{span}\{l_j(x), j \notin \Gamma_i\}, \quad i \in \Xi_1 \tag{4.13}$$

$$\mathcal{L}_{\Gamma_{\xi+1}} = span\{p_j(x), j \in \Lambda\} \tag{4.14}$$

such that $\text{span}\{l_j\} \nsubseteq [\Pi_{\Gamma_i}^*]^\perp$ for $j \in \Gamma_i, i \in \Xi_1$, and $\text{span}\{l_j\} \nsubseteq [\Pi_{\Gamma_{\xi+1}}^*]^\perp$ for $j \in \mathbf{k}$. The same result can be derived for linear systems.

In the following example, we demonstrate how to construct and generate the above set of residual signals for a given nonlinear system.

Example 4.1. Consider a nonlinear system that has 3 fault signatures and one disturbance input as governed by the following dynamics

$$\begin{aligned} \dot{x}_1 &= -x_1 x_2 + m_1 + exp(x_2) m_2 + 2m_3 + \omega_1 \\ \dot{x}_2 &= -x_1^2 - 2\frac{x_2}{x_1} m_1 + m_2 + 0.5m_3 - 0.2\omega_1 \end{aligned}$$

with the output measurement $y = [x_1, x_2]^\top$. It is clear that the above family of fault signatures does not satisfy the necessary condition (2.29), and hence it is <u>not</u> strongly detectable. Now, we show that the isolability index for the above fault signatures is 1, i.e. $\mu = 1$. First, we generate the coding sets that are required for the family of fault signatures with $\mu = 1$. Towards this end, the sets $\Gamma_i, i = 1, 2, 3$ are selected as 2 combinations of the set $\{1, 2, 3\}$, namely $\Gamma_1 = \{1, 2\}$, $\Gamma_2 = \{1, 3\}$ and $\Gamma_3 = \{2, 3\}$. The corresponding coding sets Ω_i, $i = 1, 2, 3$ are given by $\Omega_1 = \{1, 2\}$, $\Omega_2 = \{1, 3\}$ and $\Omega_3 = \{2, 3\}$ and the number of residuals is $\xi = C(3, 1) = 3$. Our next step involves checking the solvability conditions for the SFDIP problem. According to Theorem 2.6, one needs first to obtain the unobservability codistributions $\Pi^*_{\Gamma_i}, i = 1, 2, 3$. These codistributions are found by using the algorithm that is presented in [152] and are given as follows

$$\begin{aligned} \Pi^*_{\Gamma_1} &= \text{span}\{d(x_1^2 x_2)\} \\ \Pi^*_{\Gamma_2} &= \text{span}\{d(x_1 - exp(x_2))\} \\ \Pi^*_{\Gamma_3} &= \text{span}\{d(x_1 - 4x_2)\} \end{aligned}$$

It can be verified that the necessary conditions (2.27) are satisfied, and hence the isolability index for the above family of fault signatures is 1. We are now ready to design the residual generators. Towards this end, the z_1-subsystem (2.41) for each unobservability codistribution is obtained as follows

$$\Pi^*_{\Gamma_1} : \begin{cases} \dot{z}_1 = -2\frac{z_1^2}{y_2} - y_2^4 + (2\frac{z_1}{y_2} exp(\frac{z_1}{y_2^2}) + y_2^2) m_2 \\ \qquad + (4\frac{z_1}{y_2} + 0.5y_2^2) m_3 + (2\frac{z_1}{y_2} - 0.2y_2^2)\omega_1 \\ y_1 = z_1 \end{cases}$$

$$\Pi^*_{\Gamma_2} : \begin{cases} \dot{z}_1 = -(z_1 + exp(y_2)) y_2 + (1 - 2\frac{y_2}{z_1 + exp(y_2)} exp(y_2)) m_1 \\ \quad - exp(y_2)(z_1 + exp(y_2))^2 + (2 - 0.5exp(y_2)) m_3 + (1 - 0.2exp(y_2))\omega_1 \\ y_1 = z_1 \end{cases}$$

$$\Pi^*_{\Gamma_3} : \begin{cases} \dot{z}_1 = -(z_1 + 4y_2) y_2 + 4(z_1 + 4z_2)^2 + (1 + 8\frac{y_2}{z_1 + 4y_2}) m_1 \\ \qquad + (exp(y_2) - 4) m_3 + 0.2\omega_1 \\ y_1 = z_1 \end{cases}$$

The residual generators can easily be designed from the above system of equations since the state z_1 is measurable. It follows that $\Lambda_i = \Lambda =$

{1} (the disturbance input ω_1 affects all the residual signals r_i, $i = 1, 2, 3$) and according to Remark 4.2, only one extra residual signal is required. To generate this residual, one needs to find $\Pi^*_{\Gamma_4}$ that is given by

$$\Pi^*_{\Gamma_4} = \text{span}\{d(x_1 + 5x_2)\}$$

The corresponding z_1-subsystem is governed by the dynamics

$$\Pi^*_{\Gamma_4} : \begin{cases} \dot{z}_1 = -(z_1 - 5y_2)y_2 - 10(z_1 - 5y_2)^2 + (1 - 10\frac{y_2}{z_1 - 5y_2})m_1 \\ \qquad + (exp(y_2) + 5)m_2 + 4.5m_3 \\ y_1 = z_1 \end{cases}$$

Consequently, by using the above z_1-subsystem, the coding sets Ω^p_1 and Ω^f_i, $i = 1, 2, 3$ are determined as follows: $\Omega^p_1 = \{1, 2, 3\}$, $\Omega^f_1 = \{1, 2, 4\}$, $\Omega^f_2 = \{1, 3, 4\}$, and $\Omega^f_3 = \{2, 3, 4\}$. It is easy to show that these coding sets satisfy the conditions of Lemmas 4.1 and 4.2, where $\Upsilon^f_i = \{4\}, i = 1, 2, 3$.

It should be noted that since the above family of fault signatures is not strongly detectable, the method proposed in [154] cannot be applied to this system. Moreover, as a comparison with some methods in the literature that consider disturbances as faults [164, 154, 19], it should be pointed out that if the disturbance input ω_1 is treated as the fourth fault, the isolability index for the new family of fault signatures (four faults) is 1, implying that one cannot detect the concurrent occurrence of a fault and a disturbance. However, as will be shown subsequently, the proposed hybrid FDI strategy enables one to still detect a single fault in the system while a large disturbance input is applied through ω_1.

In the next section, a residual evaluation criteria is introduced for the generated residuals set $\mathfrak{R}_1 \cup \mathfrak{R}_2$.

4.2.2 Residual Evaluation Criteria

Corresponding to each residual signal $r_i \in \mathfrak{R}_1 \cup \mathfrak{R}_2$, an evaluation function is now assigned. Various evaluation functions have been introduced in Section 2.1. For the residual signals $r_i \in \mathfrak{R}_1$, two different thresholds are needed as specified below

$$J^1_{th_i} = \sup_{v \in \mathfrak{L}_p, \omega \in \mathfrak{D}_1, m_j = 0, j \in \mathbf{k}} (J_{r_i}), \quad i \in \Xi_1 \tag{4.15}$$

$$J^2_{th_i} = \sup_{v \in \mathfrak{L}_p, \omega \in \mathfrak{D}_2, m_j = 0, j \in \mathbf{k}} (J_{r_i}), \quad i \in \Xi_1 \tag{4.16}$$

In determining the first threshold, only tolerable disturbance inputs ($\omega_i \in \mathfrak{D}_1$) are considered. However, the second threshold incorporates all the possible disturbance inputs.

Remark 4.3. It should be noted that one may choose to only consider the threshold level given by $J_{th_i}^2$ as the worst case scenario associated with large disturbances. In this case, no false alarms will be generated due to disturbances. However, this leads to selection of higher threshold values that would unnecessarily reduce the sensitivity of the FDI scheme to low severity faults. As will be shown subsequently, by selecting two threshold levels and considering the temporal and sequential characteristics of the residual signals, one can not only enhance the fault sensitivity but also design and ensure a robust FDI algorithm.

The threshold values for the residual signals $r_i \in \mathfrak{R}_2$ are selected according to

$$J_{th_i}^1 = \sup_{v \in \mathfrak{L}_p, \omega_l \in \mathfrak{D}_2, i \in \Omega_l^p, m_j = 0, j \in \mathbf{k}} (J_{r_i}), \quad i = \xi + 1, ..., 2\xi \tag{4.17}$$

For a system, such as in Example 1 and the ALTAV system that is discussed in Section 4.4, where the residual signals $r_i \in \mathfrak{R}_2$ are affected by a few or even no disturbance input channels, one can select lower threshold values for these residuals. In other words, the residual signals $r_i \in \mathfrak{R}_2$ are generally less sensitive to disturbance inputs than the residual signals $r_i \in \mathfrak{R}_1$.

For each residual $r_i \in \mathfrak{R}_1$ defined at a given point in time t, we can choose the corresponding two threshold logic units $R_i^1(t)$ and $R_i^2(t)$ according to

$$R_i^1(t) = \begin{cases} 1 \text{ if } J_{r_i}(t) > J_{th_i}^1 \\ 0 \quad \text{otherwise} \end{cases}, i \in \Xi_1 \tag{4.18}$$

$$R_i^2(t) = \begin{cases} 1 \text{ if } J_{r_i}(t) > J_{th_i}^2 \\ 0 \quad \text{otherwise} \end{cases}, i \in \Xi_1 \tag{4.19}$$

Similarly, for each residual $r_i(t) \in \mathfrak{R}_2$, the threshold logic unit is assigned as follows

$$R_i^1(t) = \begin{cases} 1 \text{ if } J_{r_i}(t) > J_{th_i} \\ 0 \quad \text{otherwise} \end{cases}, i \in \{\xi + 1, ..., 2\xi\} \tag{4.20}$$

We are now in a position to classify the various fault categories that we consider in the hybrid fault diagnoser as specified in the following definition.

Definition 4.1 *The fault scenarios are categorized into the following three classes, namely, high severity faults, low severity faults, and non-detectable faults:*

1. *High severity faults correspond to faults that will affect the residual logic units* $R_i^1, i \in \{1, ..., 2\xi\}$,
2. *Low severity faults correspond to faults that will affect only* $R_i^1, i \in \{\xi + 1, ..., 2\xi\}$, *and*
3. *Non-detectable faults correspond to faults that do not affect any of the residual logic units* $R_i^1, i \in \{1, ..., 2\xi\}$.

4.2.3 DES Fault Diagnoser

For simplicity, let us assume that multiple faults in two components are possible. Furthermore, let us consider the scenario where only occurrence of one fault and one large disturbance is allowed concurrently. This assumption will limit the number of all possible operational states of the DES system. However, the proposed algorithm can easily be extended to more general cases.

First, the plant (linear/nonlinear) along with a bank of residual generators are modeled as a finite state Moore automaton [83] that is specified according to $G = (S, \Sigma, \delta, s_0, Y, \lambda)$, where S, Σ, Y are finite state, event and output sets; s_0 is the initial state, $\delta : S \times \Sigma \to S$ is the transition function, and $\lambda : S \to Y$ is the output map. For both linear and nonlinear systems, the state set S along with its description is given in Table 4.1. The event set is $\Sigma = \{\mathcal{F}_1^o, ..., \mathcal{F}_k^o, \mathcal{F}_1^r, ..., \mathcal{F}_k^r, \mathcal{D}^o, \mathcal{D}^r\}$, where the events $\mathcal{F}_i^o$ and $\mathcal{F}_i^r, i = \mathbf{k}$ correspond to the occurrence and removal of a fault in the i-th component, respectively, and the event $\mathcal{D}^o$ corresponds to the occurrence of a large disturbance in one of the $\omega_j,\ j \in \mathbf{P}$ channels and $\mathcal{D}^r$ corresponds to the removal of disturbance from all the channels. The output set is $Y = \{(R_1^1, ..., R_{\xi'}^1, R_1^2, ..., R_\xi^2) \in \mathbb{B}^\kappa\}$, where $\mathbb{B} = \{0, 1\}$ and κ can be either $2\xi + 1$ or 3ξ depending on the property of Λ_i's. Based on the above definitions, the transition function δ is now defined formally as follows

$$\begin{aligned}
&\delta(s_0, \mathcal{D}^o) = s_D, \delta(s_D, \mathcal{D}^r) = s_0, \delta(s_D, \mathcal{F}_i^o) = s_{i,D},\ i \in \mathbf{k} \\
&\delta(s_0, \mathcal{F}_i^o) = s_i, \delta(s_i, \mathcal{D}^o) = s_{i,D}, \delta(s_i, \mathcal{F}_i^r) = s_0,\ i \in \mathbf{k} \\
&\delta(s_i, \mathcal{F}_j^o) = s_{i,j}, \delta(s_{i,j}, \mathcal{F}_i^r) = s_j, \delta(s_{i,j}, \mathcal{F}_j^r) = s_i,\ i, j \in \mathbf{k}, i \neq j \\
&\delta(s_{i,D}, \mathcal{F}_i^r) = s_D, \quad \delta(s_{i,D}, \mathcal{D}^r) = s_i,\ i \in \mathbf{k}
\end{aligned}$$

As an illustration, Figure 4.2 shows the corresponding transition functions of the nonlinear system that are considered in Example 4.1. It should be noted that since for this system the isolability index is $\mu = 1$, multiple fault states are not applicable. As shown in Figure 4.2, the DES model of the integrated nonlinear system and the bank of residual generators in Example 4.1 have eight states, namely the normal operational state s_0, three faulty states $s_i, i = 1, 2, 3$, three concurrent fault and large disturbance states $s_{i,D}, i = 1, 2, 3$, and the large disturbance input s_D. The event set and the output set for this system are $\Sigma = \{\mathcal{F}_1^o, \mathcal{F}_2^o, \mathcal{F}_3^o, \mathcal{F}_1^r, \mathcal{F}_2^r, \mathcal{F}_3^r, \mathcal{D}^o, \mathcal{D}^r\}$ and $Y = \{(R_1^1, ..., R_4^1, R_1^2, ..., R_3^2) \in \mathbb{B}^7\}$, respectively.

The output map λ depends on the severity of a fault and the threshold values for the residual signals. As mentioned in the previous section, threshold values for the residual signals $r_i \in \mathfrak{R}_2$ are usually lower than those of $r_1 \in \mathfrak{R}_1$. Therefore, there could be a low severity fault scenario where the residual logic unit $R_{\xi+i}^1$ becomes one while R_i^1 is zero. In defining the output map λ, such scenarios are also incorporated. Table 4.2 shows the corresponding output

Table 4.1 Finite states of the plant

	Operational state	Description
s_0	Normal operation	No fault and no large disturbance input
s_1	Fault state 1	Fault in the first actuator
$\vdots$	$\vdots$	$\vdots$
s_k	Fault state k	Fault in the k-th actuator
$s_{1,2}$	Multiple faults state (1,2)	Faults in the first and the second actuators
$\vdots$	$\vdots$	$\vdots$
$s_{(k-1),k}$	Multiple faults state (k-1,k)	Faults in the $(k-1)$ and the k-th actuators
s_D	Disturbance state	Occurrence of large disturbance input
$s_{1,D}$	Fault/Disturbance state 1	Fault in the first actuator and large disturbance inputs
$\vdots$	$\vdots$	$\vdots$
$s_{k,D}$	Fault/Disturbance state k	Fault in the k-th actuator and large disturbance inputs

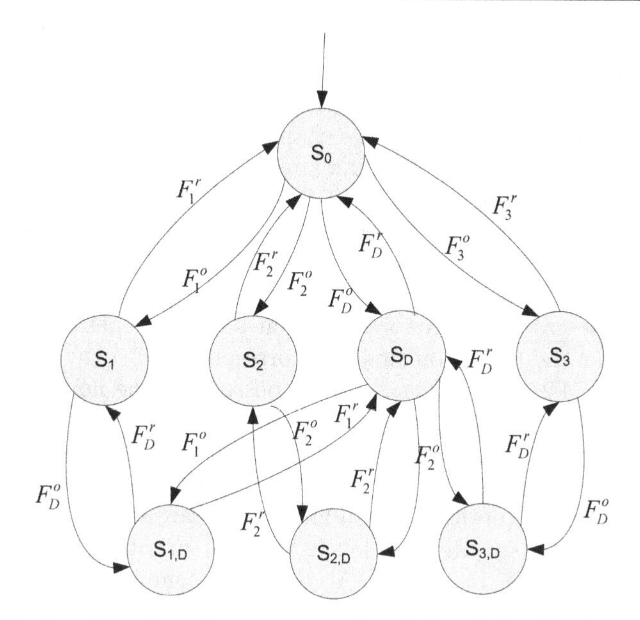

Fig. 4.2 The transition functions corresponding to Example 1 [135].

map λ where $\Xi_2 = \{1, ..., 2\xi\}$. Thus, some states may have different outputs that would depend on the severity of the fault and the disturbances. Moreover, non-detectable fault scenarios (refer to Definition 4.1) are not observable from the residual logic units, and therefore they cannot be detected and isolated. These types of faults are not considered in λ. In other words, no event is assigned to such faults.

Table 4.2 Output maps of the plant

	Output map λ
s_0	$(0, ..., 0)$
s_D	$\{(R_1^1, ..., R_\xi^2) \in Y \mid \exists l \in \mathbf{P}, \forall j \in \Omega_l^p, R_j^1 = 1\}$
s_i	$\{(R_1^1, ..., R_\xi^2) \in Y \mid \exists \beta \in \{1,2\}, j \in \Omega_i^f, R_j^\beta = 1\}$
$s_{i,j}$	$\{(R_1^1, ..., R_\xi^2) \in Y \mid \exists \beta \in \{1,2\}, k \in \Omega_i^f \cup \Omega_j^f,\ R_k^\beta = 1\}$
$s_{i,D}$	$\{(R_1^1, ..., R_\xi^2) \in Y \mid \exists \beta \in \{1,2\}, j \in \Omega_i^f, R_j^\beta = 1\}$ $\cup \{(R_1^1, ..., R_\xi^2) \in Y \mid \exists l \in \mathbf{P}, \forall j \in \Omega_l^p, R_j^1 = 1\}$

The purpose of the DES diagnoser is to utilize the output sequence of the system (residual logic units) as inputs and to generate an estimate of the state of the system. In this chapter, a DES diagnoser is modeled as a finite state automaton $H = (S_H, I_H, \delta_H, z_0, Y_H, \lambda_H)$, where S_H, I_H, Y_H denote the finite state, input and output sets; z_0 is the initial state of the diagnoser, $\delta_H : S_H \times I_H \rightarrow S_H$ denotes the transition function, and λ_H is the output map. In order to eliminate any possible ambiguity in the DES model (G) output, two additional states with respect to the state set of G are considered for H, namely $S_H = \{S, s_F, s_{F,D}\}$, where s_F corresponds to the faulty state where one cannot isolate the faulty channel and $s_{F,D}$ corresponds to the concurrent occurrence of a fault and a large disturbance in the system when a fault may not be isolated. The input set for the diagnoser is an output set of G (set Y). The output set is the same as the state set of the diagnoser ($Y_H = S_H$) and the output map $\lambda_H : S_H \rightarrow Y_H$ is an identity map.

The main step that is left is the design of a transition map δ_H. First, we consider the case when the system is in a normal operational mode s_0 and try to find the transition function corresponding to this mode. Based on Assumption 4.2, three transitions are possible in the normal operation, namely transition to the state s_i which corresponds to the occurrence of a fault in the i-th actuator (event $\mathcal{F}_i^o$), transition to the state s_D which corresponds to the occurrence of a large disturbance in one of the input disturbance channels (event $\mathcal{D}^o$), and finally the transition to the fault mode s_F which corresponds to the occurrence of a low severity fault in one of the actuators that may not be isolable. According to Lemma 4.1, the effects of a fault and a disturbance can be distinguished easily by the fault diagnoser by using the coding sets Ω_i^f and Ω_j^p, and therefore the sets Ω_i^f and Ω_j^p can be used for the transition to states s_i and s_D, respectively.

The only remaining case of interest is when the occurrence of a low severity fault (refer to Definition 4.1) in the i-th component will lead to changes in only $r_j \in \mathfrak{R}_2$. In this case, according to Lemma 4.2, the diagnoser can *detect* the occurrence of a fault in the system since $\Upsilon_i^f \neq \Omega_j^p$. However, we may have $\Upsilon_i^f = \Upsilon_j^f$ for some $i, j \in \mathbf{k}$, and therefore the fault cannot be *isolated.* In this case, the state of the fault diagnoser will change to s_F. Table 4.3 summarizes the transition function that is initiated from the state s_0 where all the unspecified residual logic units are zero.

Table 4.3 Transition functions of state s_0

Current State	Input $(R_1^1, ..., R_{2\xi}^1, R_1^2, ..., R_\xi^2)$	Next state	Corresponding event in G
s_0	$\bigwedge_{j\in\Omega_i^f} R_j^1 = 1$	$s_i, i \in \mathbf{k}$	$\mathcal{F}_i^o$
s_0	$\exists l \in \mathbf{k}$ such that $\bigwedge_{j\in\Upsilon_l^f} R_j^1 = 1$	s_F	$\mathcal{F}_i^o \in \{\mathcal{F}_1^o, ..., \mathcal{F}_k^o\}$
s_0	$\exists l \in \mathbf{P}$ such that $\bigwedge_{j\in\Omega_l^p} R_j^1 = 1$	s_D	$\mathcal{D}_i^o$

The next step is now to consider scenarios when initially a large disturbance is applied to the system followed by a fault that is concurrently present in one of the system actuators. Therefore, it is assumed that the system has a transition from the normal operation state s_0 to the disturbance state s_D where we define a set $\mathcal{D} = \{1 \leq i \leq \xi | R_i^1 = 1\}$. In this state, the second threshold logic units R_i^2 are used for all the residuals $r_i, i \in \mathcal{D}$. The transition function for the state s_D is given in Table 4.4. It is assumed that the effects of the fault is not nullified by a large disturbance input, which is quite a reasonable consideration for practically most situations.

Table 4.4 Transition functions of state s_D

Current state	Input $(R_1^1, ..., R_{2\xi}^1, R_1^2, ..., R_\xi^2)$	Next state	Corresponding event in G
s_D	all inputs become zero	s_0	$\mathcal{D}^r$
s_D	$(\bigwedge_{j\in\Omega_i^f\cup\mathcal{D}} R_j^2 \bigwedge_{l\in\Upsilon_i^f} R_l^1 = 1)$	$s_{i,D}$	$\mathcal{F}_i^o$
s_D	$\exists l \in \mathbf{k}$ and $\exists j \in \mathbf{P}$ such that $\bigwedge_{i\in\Upsilon_l^f\cup\Omega_j^p} R_i^1 = 1$	$s_{F,D}$	$\mathcal{F}_l^o \in \{\mathcal{F}_1^o, ..., \mathcal{F}_k^o\}$

Now let us consider a scenario where a fault is detected in the i-th actuator and the state of the fault diagnoser is s_i. Generally, we should investigate three possible cases, namely 1) the removal of a detected fault, 2) the occurrence of a second fault in the j-th actuator, and 3) the occurrence of a disturbance in $\omega_l, l \in \mathbf{P}$. Actually, the main challenge here is to distinguish between cases 2 and 3, since the removal of a fault can be easily detected when all the threshold logic units become zero. The necessary condition for distinguishing between cases 2 and 3 is governed by

$$\Omega_i^f \cup \Omega_j^f \neq \Omega_i^f \cup \Omega_l^p, \; i, j \in \mathbf{k}, ; l \in \mathbf{P} \tag{4.21}$$

The next lemma provides the sufficient condition for satisfying the condition (4.21).

Lemma 4.3 *If the number of residuals $r_i \in \mathfrak{R}_1$ that are affected by each disturbance input is more than $|\Omega_i \cup \Omega_j| = \xi - C(k-2, k-\mu)$, i.e. $|\Omega_l^p \cap \Xi_1| > \xi - C(k-2, k-\mu), \forall l \in \boldsymbol{P}$, then the condition (4.21) is satisfied for all the disturbance inputs $l \in \boldsymbol{P}$ as well as fault modes $m_i, m_j, i \neq j$.*

Proof: If $|\Omega_l^p \cap \Xi_1| > \xi - C(k-2, k-\mu)$, then for any two fault modes $m_i, m_j, i, j \in \mathbf{k}, i \neq j$, there exists at least one residual $r_\alpha \in \mathfrak{R}_1$ such that $r_\alpha \in \Omega_l^p$ and $r_\alpha \notin \Omega_i^f \cup \Omega_j^f$, and therefore it follows that $\Omega_i^f \cup \Omega_j^f \neq \Omega_i^f \cup \Omega_l^p$. ■

Table 4.5 Transition functions of state $s_i,\ i \in \mathbf{k}$

Current state	Input $(R_1^1, ..., R_{2\xi}^1, R_1^2, ..., R_\xi^2)$	Next state	Corresponding event in G
s_i	all inputs become zero	s_0	$\mathcal{F}_i^r$
s_i	$\bigwedge_{l\in\Omega_i^f\cup\Omega_j^f} R_l^1 = 1$ for the time interval τ_0	$s_{i,j}$	$\mathcal{F}_j^o$
s_i	$\exists l$ such that $\bigwedge_{j\in\Omega_i^f\cup\Omega_l^p} R_j^1 = 1$	$s_{i,D}$	$\mathcal{D}^o$

It can be easily verified that the system in Remark 4.2 where $\Lambda_i = \Lambda, i \in \mathbf{k}$ satisfies the above sufficient condition if $\mu > 1$, since $|\Omega_l^p \cap \Xi_1| = \xi$ and $|\Omega_i^f \cup \Omega_j^f| < \xi$.

Remark 4.4. In a situation where $\Omega_i^f \cup \Omega_j^f \subset \Omega_i^f \cup \Omega_l^p$, one could potentially have a false alarm associated with the second fault while a large disturbance input is present. To remedy this problem, the DES diagnoser will declare the detection of the second fault after a specific waiting-time interval of τ_0 seconds, if all the residual threshold logics specified by $\Omega_i^f \cup \Omega_j^f$ are at 1 while the remaining residual threshold logic units specified by $\{\Omega_i^f \cup \Omega_j^f\} - \{\Omega_i^f \cup \Omega_l^p\}$ remain at zero. Table 4.5 illustrates the transition functions for the states associated with $s_i,\ i \in \mathbf{k}$.

Table 4.6 shows the remaining transitions that should be considered for the DES diagnoser. By specifying these transitions, the design of the hybrid DES diagnoser is completed.

Table 4.6 Transition functions of states s_F, $s_{F,D}$, $s_{i,D}$ and $s_{i,j}$

Current state	Input $(R_1^1, ..., R_{2\xi}^1, R_1^2, ..., R_\xi^2)$	Next state	Corresponding event in G
s_F	$\bigwedge_{j\in\Omega_i^f} R_j^1 = 1$ for the time interval τ_0	s_i	$\mathcal{F}_i^o$
s_F	$\exists l \in \mathbf{k}$ and $\exists j \in \mathbf{P}$ such that $\bigwedge_{i\in\Upsilon_l^f\cup\Omega_j^p} R_i^1 = 1$	$s_{F,D}$	$\mathcal{D}^o$
$s_{F,D}$	$\bigwedge_{j\in\Omega_i^f\cup\mathcal{D}} R_j^2 \bigwedge_{l\in\Upsilon_i^f} R_l^1 = 1$	$s_{i,D}$	$\mathcal{F}_i^o$
$s_{F,D}$	$\exists l \in \mathbf{k}$ such that $\bigwedge_{j\in\Upsilon_l^f} R_j^1 = 1$	s_F	$\mathcal{D}^r$
$s_{F,D}$	$\exists l \in \mathbf{P},\ i \in \mathbf{k}$ such that $\bigwedge_{j\in\Omega_l^p} R_j^1 = 1$	s_D	$\mathcal{F}_i^r \in \{\mathcal{F}_1^o, ..., \mathcal{F}_k^o\}$
$s_{i,D}$	$\exists l \in \mathbf{P}$ such that $\bigwedge_{j\in\Omega_l^p} R_j^1 = 1$	s_D	$\mathcal{F}_i^r \in \{\mathcal{F}_1^o, ..., \mathcal{F}_k^o\}$
$s_{i,D}$	$\bigwedge_{j\in\Omega_i^f} R_j^1 = 1$	s_i	$\mathcal{D}^r$
$s_{i,j}$	$\bigwedge_{l\in\Omega_i^f} R_l^1 = 1$	s_i	$\mathcal{F}_j^r$
$s_{i,j}$	$\bigwedge_{l\in\Omega_j^f} R_l^1 = 1$	s_j	$\mathcal{F}_i^r$

***Example 4.1** (Cont.)* According to the coding sets that are obtained for the nonlinear system in Example 4.1, the DES diagnoser can be designed as follows: the state set is specified by $S_H = \{s_0, s_1, s_2, s_3, s_D, s_F, s_{1,D}, s_{2,D}, s_{3,D}\}$, the input set is defined by $I_H = \{R_1^1, R_2^1, R_3^1, R_4^1, R_1^2, R_2^2, R_3^2\}$ and the transition map λ_H is given in Table 4.7. Therefore, the design of the proposed hybrid diagnoser for the nonlinear system in Example 4.1 is completed.

Table 4.7 Transition functions of Example 4.1

Current State	Input $(R_1^1, ..., R_4^1, R_1^2, ..., R_3^2)$	Next state
s_0	$R_1^1 \bigwedge R_2^1 \bigwedge R_4^1 = 1$	s_1
s_0	$R_1^1 \bigwedge R_3^1 \bigwedge R_4^1 = 1$	s_2
s_0	$R_2^1 \bigwedge R_3^1 \bigwedge R_4^1 = 1$	s_3
s_0	$R_1^1 \bigwedge R_2^1 \bigwedge R_3^1 = 1$	s_D
s_0	$R_4^1 = 1$	s_F
s_i	all zero	s_0
s_D	all zero	s_0
s_D	$R_1^2 \bigwedge R_2^2 \bigwedge R_4^1 = 1$	$s_{1,D}$
s_D	$R_1^2 \bigwedge R_3^2 \bigwedge R_4^1 = 1$	$s_{2,D}$
s_D	$R_2^2 \bigwedge R_3^2 \bigwedge R_4^1 = 1$	$s_{3,D}$
s_F	all zero	s_0
s_F	$R_1^1 \bigwedge R_2^1 \bigwedge R_4^1 = 1$	s_1
s_F	$R_1^1 \bigwedge R_3^1 \bigwedge R_4^1 = 1$	s_2
s_F	$R_2^1 \bigwedge R_3^1 \bigwedge R_4^1 = 1$	s_3
s_F	$R_1^1 \bigwedge R_2^1 \bigwedge R_3^1 \bigwedge R_4^1 = 1$	$s_{F,D}$

In the next two sections, the proposed hybrid FDI strategy is applied to the actuator FDI problem in a network of unmanned vehicles (linear systems) and the ATLAV system (nonlinear systems).

4.3 Hybrid Actuator FDI in a Network of Unmanned Vehicles

In this section, the hybrid FDI method is now applied to the actuator fault detection and isolation (FDI) problem for a network of N vehicles whose (identical or homogenous) linear dynamics are governed by

$$\dot{x}_i(t) = Ax_i(t) + Bu_i(t) + \sum_{k=1}^{a} L_k m_{ik}(t) + \sum_{j=1}^{q} P_j \omega_{ij}(t) \tag{4.22}$$

where the fault signature L_k represents a fault in the k-th actuator of the vehicle, i.e, L_k is the k-th column of B. It is assumed that matrix B is full rank (Rank(B)=a). It should be noted that in comparison with the vehicle dynamics considered in Section 3.3, additional disturbance terms are added to the vehicle dynamics. Similar relative state measurements are considered for each vehicle as

$$z_{ij}(t) = C(x_i(t) - x_j(t)) \qquad j \in \mathrm{N}_i \tag{4.23}$$

where the set $\mathrm{N}_i \subset [1, N]\backslash i$ represents the set of vehicles that vehicle i can sense and is designated as the neighboring set of the vehicle i, and $z_{ij} \in \mathcal{Z}_i, j \in \mathrm{N}_i$ represents the state measurement relative to the other vehicles. It is assumed that the pair (A, C) is observable.

Assumption 4.4 *It is assumed that the family of fault signature L_i's are strongly detectable when each vehicle has an absolute state measurement $y_i = Cx_i$.*

In this section only semi-decentralized FDI architecture (Section 3.3.2) is investigated, and hence it is assumed that local communication links exist between each vehicle and its neighbors and the control signals u_i are communicated among the vehicles.

Let $z_i = [z_{ii_1}^\top, z_{ii_2}^\top, \cdots, z_{ii_{|\mathrm{N}_i|}}^\top]^\top$ and $\mathrm{N}_i = \{i_1, i_2, ..., i_{|\mathrm{N}_i|}\}$. Since the output measurement z_i depends on the state of the neighboring vehicles, the following *nodal model* should be considered for the i-th vehicle in order to design an FDI filter, namely

$$\begin{aligned}\dot{x}_{\mathrm{N}_i}(t) &= A^{|\mathrm{N}_i|+1} x_{\mathrm{N}_i}(t) + B^{|\mathrm{N}_i|+1} u_{\mathrm{N}_i}(t) \\ &+ \sum_{k=0}^{|\mathrm{N}_i|} \sum_{l=1}^{a} \bar{L}_{(k+1)l} m_{i_k l}(t) + \sum_{k=0}^{|\mathrm{N}_i|} \sum_{j=1}^{q} \bar{P}_{(k+1)j} \omega_{i_k j}(t)\end{aligned} \tag{4.24}$$

with measurement $z_i(t) = \bar{C}_{|\mathrm{N}_i|} x_{\mathrm{N}_i}(t)$, where $x_{\mathrm{N}_i} = [x_i^\top, x_{i_1}^\top, \cdots, x_{i_{|\mathrm{N}_i|}}^\top]^\top$, $u_{\mathrm{N}_i} = [u_i^\top, u_{i_1}^\top, \cdots, u_{i_{|\mathrm{N}_i|}}^\top]^\top$, $\bar{L}_{(k+1)l}$ is the $k \times a + l$ column of $B^{|\mathrm{N}_i|+1}$, $\bar{P}_{(k+1)j}$ is the $k \times d + j$ column of matrix $I_{|\mathrm{N}_i|+1} \otimes P$, $P = [P_1, ..., P_q]$, and $\bar{C}_{|\mathrm{N}_i|}$ is constructed as in (3.7) from matrix C. Moreover, we define $m_{i_0 l}$ and $\omega_{i_0 j}$ as the fault signatures and disturbance inputs corresponding to the i-th vehicle, i.e. m_{il} and ω_{ij}, respectively. Therefore, $\bar{L}_{1l}$ represents the fault signatures of the l-th actuator of the i-th vehicle in the nodal system (4.24).

It follows that the entire state x_{N_i} is not fully observable from the relative state measurements z_i, and the states of the centroid of vehicle i and its neighbors cannot be determined from z_i. Since the pair $(A^{|\mathrm{N}_i|+1}, \bar{C}_{|\mathrm{N}_i|})$ is not observable, one can first try to obtain the observable part of system (4.24) as

$$\dot{x}_{\mathrm{N}_i}^O(t) = A^{|\mathrm{N}_i|} x_{\mathrm{N}_i}^O(t) + \bar{B}_{|\mathrm{N}_i|}^O u_{\mathrm{N}_i}(t) + \sum_{l=1}^{K} \bar{L}_l^O \bar{m}_l(t) + \sum_{j=1}^{Q} \bar{P}_j^O \bar{\omega}_j(t) \tag{4.25}$$

with measurement $z_i(t) = C^{|\mathrm{N}_i|} x^O_{\mathrm{N}_i}(t)$, where $x^O_{\mathrm{N}_i} = [x_i^\top - x_{i_1}^\top, x_i^\top - x_{i_2}^\top, \cdots, x_i^\top - x_{i_{|\mathrm{N}_i|}}^\top]^\top \in \mathbb{R}^{n|\mathcal{N}_i|}$. To avoid notational complexity by having double indices, the terms $m_{i_k l}(t)$ and $\omega_{i_k j}(t)$ are re-written as $\bar{m}_l(t)$ and $\bar{\omega}_j(t)$ where $K = a \times (|\mathrm{N}_i| + 1)$, $Q = q \times (|\mathrm{N}_i| + 1)$, $\bar{m}_l(t)$ with $l = 1, ..., a$ corresponds to the fault signals of the i-th vehicle, namely $m_{i_0 l}(t)$'s, and $\bar{m}_l(t)$ with $l = a+1, ..., 2a$ corresponds to the fault signals of the first neighbor of the vehicle i, namely, $m_{i_1 l}(t)$'s, etc. A similar association is considered between $\bar{\omega}_j(t)$'s and $\omega_{i_k j}(t)$'s. The fault signatures $\bar{L}^O_l$ and the disturbance signatures $\bar{P}^O_j$ also have a one-to-one association with $\bar{L}^O_{(k+1)l}$ and $\bar{P}^O_{(k+1)j}$, respectively, where $\bar{L}^O_{(k+1)l}$ is the $(ka+l)$-th column of $\bar{B}^O_{|\mathrm{N}_i|}$ and $\bar{P}^O_{(k+1)j}$ is the $(ka+j)$-th column of $\bar{P}^O_{|\mathrm{N}_i|}$ and $\bar{B}^O_{|\mathrm{N}_i|}$ and $\bar{P}^O_{|\mathrm{N}_i|}$ are constructed as in (3.7) from matrices B and P, respectively.

As discussed in Section 3.3.2, the family of $a \times (|\mathrm{N}_i| + 1)$ fault signatures $\bar{L}^O_{kj}$, $k = 1, ..., |\mathrm{N}_i|+1$, $j = 1, ..., a$ satisfies the conditions of Theorem 3.6 with the isolability index of $\mu = |\mathrm{N}_i| - 1$. Based on Assumption 4.4, the SFDIP problem has a solution for the coding scheme 2 and one needs to generate $C(a \times (|\mathrm{N}_i|+1), |\mathrm{N}_i|-1)$ residuals for detecting and isolating $|\mathrm{N}_i|-1$ multiple faults.

As pointed out in Remark 4.1, considering the disturbance inputs as faults in the system limits the number of faults that one can then detect and isolate, i.e. , it will decrease the isolability index of fault signatures. However, by using the proposed FDI algorithm, one can keep the isolability index of the fault signatures as high as possible.

4.3.1 Simulation Results

In this section, the proposed hybrid FDI algorithm is applied to a network of four quad-rotors. The quad-rotor dynamics are assumed to be governed by the following equations [29]

$$\begin{aligned} m\ddot{x} &= -usin(\theta) \\ m\ddot{y} &= ucos(\theta)sin(\phi) \\ m\ddot{z} &= ucos(\theta)cos(\phi) - mg \\ \ddot{\psi} &= \tilde{\tau}_\psi, \; \ddot{\theta} = \tilde{\tau}_\theta, \; \ddot{\phi} = \tilde{\tau}_\phi \end{aligned} \tag{4.26}$$

where x and y are the coordinates in the horizonal plane, z is the vertical position, ψ is the yaw angle about the z-axis, θ is the pitch angle about the y-axis, and ϕ is the roll angle about the x-axis. Each rotor generates the thrust $f_i = k_i \omega_i^2$, where $k_i > 0$ is a constant and ω_i is the angular speed of motor i. The control input u is the total thrust or collective input which is obtained as $u = f_1 + f_2 + f_3 + f_4$. The angular moments about each axis are as follows: $\tau_\psi = \sum_{i=1}^4 \tau_{M_i}$, $\tau_\theta = (f_2 - f_4)\tilde{l}$, and $\tau_\phi = (f_3 - f_1)\tilde{l}$, where $\tilde{l}$ is the distance measured from the motors to the center of gravity and τ_{M_i} is the coupling produced by the motors.

The new angular momentums $\tilde{\tau}_\psi$, $\tilde{\tau}_\theta$ and $\tilde{\tau}_\phi$ are defined based on the change of input variable $\tau = C(\eta, \dot{\eta})\dot{\eta} + \mathbb{J}\tilde{\tau}$ [29], where $\tilde{\tau} = [\tilde{\tau}_\psi, \tilde{\tau}_\theta, \tilde{\tau}_\phi]^\top$, $\tau = [\tau_\psi, \tau_\theta, \tau_\phi]^\top$, $\eta = [\psi, \theta, \phi]^\top$, $C(\eta, \dot{\eta})$ is the Coriolis term and contains the gyroscopic and centrifugal forces associated with the η dependence of $\mathbb{J}$, and $\mathbb{J}$ is the inertia matrix of the full-rotational kinetic energy of the quad-rotor expressed in terms of η.

For design of the residual generators it is assumed that the quad-rotor is moving in a plane with fixed altitude and the Euler angles ψ, θ and ϕ are sufficiently small. Hence, we have $u \approx mg$ and equation (4.26) can be re-written as follows

$$\begin{aligned}
\ddot{x}(t) &= -g\theta(t) + A_w(t) \\
\ddot{\theta}(t) &= \tilde{\tau}_\theta(t) + \tilde{m}_\theta(t) \\
\ddot{y}(t) &= g\phi(t) + A_w(t) \\
\ddot{\phi}(t) &= \tilde{\tau}_\phi(t) + \tilde{m}_\phi(t)
\end{aligned} \tag{4.27}$$

where we have used the approximations $sin(\theta) = \theta$, $sin(\phi) = \phi$, and $cos(\theta) = 1$. The term A_w that is added to the x and y dynamics is the resulting aerodynamic force due to the wind turbulence and wind gust. This force can be computed from the aerodynamic coefficients C_i as $A_i = \frac{1}{2}\rho_{air} C_i V^2$ [17], where ρ_{air} is the air density, and V is the velocity of the quad-rotor with respect to air, i.e. $V = V_r - V_{wind}$, where $V_r = \sqrt{\dot{x}^2 + \dot{y}^2}$ is the velocity of the quad-rotor.

The terms $\tilde{m}_\theta(t)$ and $\tilde{m}_\phi(t)$ represent the fault signals corresponding to the angular momentums $\tilde{\tau}_\theta(t)$ and $\tilde{\tau}_\phi(t)$, respectively. It follows that $\tilde{m}_\theta(t)$ and $\tilde{m}_\phi(t)$ correspond to the faults in motors 2 and 4, and motors 1 and 3, respectively. It should be noted that the z dynamics as well as the yaw angle ψ dynamics are ignored given the fact that the quad-rotor is in a hovering mode.

It is assumed that the wind is generated by two different phenomena, namely, wind turbulence and wind gust. The Dryden model [7] is used here for generating a wind turbulence signal through the relationship $V_{turb} = G_w\varepsilon$, where ε is a zero-mean Gaussian white noise and V_{turb} is the random vertical wind turbulence process. The Dryden filter $G_w(s)$ is expressed as: $G_w(s) = \sqrt{\frac{3U_0\sigma_w^2}{\pi L_w}\frac{s+\frac{U_0}{\sqrt{3}L_w}}{[s+\frac{U_0}{L_w}]^2}}$, where σ_w is the RMS vertical wind velocity (m/s), L_w is the scale for the disturbance, and U_0 is the vehicle trim velocity (m/s). Moreover, a discrete wind gust model is considered here based on the Military Specification MIL-F-8785C [6]. The mathematical representation of the discrete gust is given by

$$V_{gust} = \begin{cases} 0 & x < 0 \\ \frac{V_m}{2}(1 - cos(\frac{\pi x}{d_m})) & 0 \leq x \leq d_m \\ V_m & x > d_m \end{cases} \tag{4.28}$$

where V_m is the wind gust amplitude, d_m is the gust length, x is the distance traveled, and V_{gust} is the resultant wind gust velocity. Therefore, the total wind speed is the sum of the wind gust V_{gust} and the wind turbulence V_{turb}.

The neighboring sets for the network studied here are given by $\mathrm{N}_1 = \{2, 3\}$, $\mathrm{N}_2 = \{1\}$, $\mathrm{N}_3 = \{4\}$ and $\mathrm{N}_4 = \{1\}$. Without loss of generality, we only consider the actuator FDI problem for the quad-rotor 1 with $i_1 = 2$ and $i_2 = 3$ since $\mathrm{N}_1 = \{2, 3\}$. It is assumed that the four quad-rotors are maintaining a formation while following the trajectory of the first quad-rotor. The reference trajectory for the first vehicle is considered to be $x = y = t$ where t denotes the simulation time.

The isolability index for the family of actuator fault signatures in the nodal model of the quad-rotor 1 is 1, i.e. $\mu = |\mathcal{N}_1| - 1 = 1$. The fault signatures $\bar{L}_l^O$'s can be categorized into two subsets, namely, $FL_1 = \{\bar{L}_1^O, \bar{L}_3^O, \bar{L}_5^O\}$ and $FL_2 = \{\bar{L}_2^O, \bar{L}_4^O, \bar{L}_6^O\}$, and the sets Γ_k^1 and Γ_k^2 are selected as the 2 combinations of the index sets $I_1 = \{1, 3, 5\}$ and $I_2 = \{2, 4, 6\}$. Hence, six residuals $r_1, ..., r_6$ are needed for the residual set $\mathfrak{R}_1$, where $\Gamma_1^1 = \{3, 5\}$, $\Gamma_2^1 = \{1, 5\}$, $\Gamma_3^1 = \{1, 3\}$, $\Gamma_1^2 = \{4, 6\}$, $\Gamma_2^2 = \{2, 6\}$ and $\Gamma_3^2 = \{2, 4\}$. The set Γ_k's are defined as $\Gamma_k = \Gamma_k^1$ for $k = 1, 2, 3$, and $\Gamma_k = \Gamma_k^2$ for $k = 4, 5, 6$.

We have $\Lambda_1 = \Lambda_4 = \{\bar{\omega}_2, \bar{\omega}_3\}$, $\Lambda_2 = \Lambda_5 = \{\bar{\omega}_1, \bar{\omega}_3\}$, and $\Lambda_3 = \Lambda_6 = \{\bar{\omega}_1, \bar{\omega}_2\}$, where $\bar{\omega}_j$ denotes the wind disturbance in the j-th quad-rotor, i.e. $\bar{\omega}_j = \omega_{j1}$ $(q = 1)$. However, due to the structure of $\bar{P}^O_{|\mathrm{N}_i|}$, we have $\bar{\mathcal{P}}_j^O \in \mathcal{S}^*_{\xi+k}$ for all $k \in \Xi$ and $j = 1, 2, 3$. In other words, all the residuals in the complementary set $\mathfrak{R}_2$ are decoupled from all disturbances. It can also be verified that these residuals are affected by all actuator faults $\bar{m}_l$ in the nodal system (4.25) for the quad-rotor 1. Therefore, these residuals are identical and only one extra residual, namely r_7 is needed for this system which is decoupled from all disturbance inputs $\bar{d}_1$, $\bar{d}_2$ and $\bar{d}_3$ and is affected by all the actuator faults of the quad-rotors 1, 2, and 3.

Consequently, we have $\Omega_1^p = \{2, 3, 5, 6\}$, $\Omega_2^p = \{1, 3, 4, 6\}$, $\Omega_3^p = \{1, 2, 4, 5\}$, $\Omega_1^f = \{2, 3, 7\}$, $\Omega_2^f = \{5, 6, 7\}$, $\Omega_3^f = \{1, 3, 7\}$, $\Omega_4^f = \{4, 6, 7\}$, $\Omega_5^f = \{1, 2, 7\}$, and $\Omega_6^f = \{4, 5, 7\}$ and $\Upsilon_l^f = \{7\}$ for $l = 1, ..., 6$ where Ω_1^f and Ω_2^f correspond to the first and the second actuator faults of the quad-rotor 1, Ω_3^f and Ω_4^f correspond to the first and the second actuator faults of the quad-rotor 2, and Ω_5^f and Ω_6^f correspond to the first and the second actuator faults of the quad-rotor 3.

The evaluation functions for the residuals r_k are selected as $J_{r_k}(t) = \frac{1}{T_0}\int_{t-T_0}^{t} r_k^\top(t) r_k(t) dt, k = 1, ..., 7$, where $T_0 = 5$ seconds is the length of the evaluation window. For conducting simulations, a typical value for the wind turbulence is selected as $L_w = 580m$, $\sigma_w = 7m/s$, and $U_0 = 1$. It is assumed that the wind gust exists between $x = 50$m and $x = 100$m with parameters $d_m = 10m$ and $V_m = 8m/s$. By conducting Monte Carlo simulations and considering (a) the worst case scenario of the residuals corresponding to the healthy mode of the network subject to measurement noise with uniform distribution of $\pm$ 0.001, and (b) the wind turbulence and wind gust, the threshold values are selected as $J^1_{th_k} = 0.0035, J^2_{th_k} = 0.02,\ k = 1, ..., 6$, and $J^1_{th_7} = 0.0006$. It should be pointed out that since the residual r_7 is decoupled from the disturbance inputs, one can select a lower threshold value for it.

The next step is to design the DES fault diagnoser H. There exist six fault signals $\bar{m}_l, l = 1, ..., 6$, and hence the state modes $s_l, l = 1, ..., 6$, are assigned to the occurrence of a single fault where s_1, s_3 and s_5 correspond to faults in the first actuator of the quad-rotors 1, 2, and 3, respectively, and s_2, s_4 and s_6 correspond to faults in the second actuator of the quad-rotors 1, 2, and 3, respectively. One can also detect and isolate concurrent faults in the first actuator of the quad-rotor i and the second actuator of the quad-rotor j for all $i, j = 1, 2, 3$. Hence, the states $s_{1,2}$, $s_{1,4}$, $s_{1,6}$, $s_{2,3}$, $s_{2,5}$, $s_{3,4}$, $s_{3,6}$, $s_{4,5}$, and $s_{5,6}$ are assigned to the occurrence of concurrent faults, where $s_{i,j}$ corresponds to concurrent occurrence of faults specified by s_i and s_j. It should be noted that the states $s_{1,3}$, $s_{1,5}$, $s_{3,5}$, $s_{2,4}$, $s_{2,6}$ and $s_{4,6}$ are not considered since they correspond to the concurrent occurrence of faults in the first actuator of the two quad-rotors or the concurrent occurrence of faults in the second actuator of the two quad-rotors. These faults cannot be isolated by using the coding scheme since the isolability index for the fault signatures in the nodal system of the quad-rotor 1 is $\mu = 1$.

The states $s_{l,D}$ refer to the occurrence of concurrent fault s_l and a large disturbance. Consequently, the cardinality of the state set of the quad-rotor 1 is 23. The input set of the diagnoser is $I = \{(R^1_1, ..., R^1_7, R^2_1, ..., R^2_6) \in \mathbb{B}^{13}\}$, and the output set is equal to S_H. The transition function λ_H can be found by following the results in Section 4.2.3.

Figures 4.3 and 4.4 show the residual evaluation functions and the state of the HFD corresponding to a concurrent fault scenario where the fault $\bar{m}_3 = m_{21} = -0.015$ is injected in the quad-rotor 2 at $t = 120$ seconds ($\tilde{m}_\theta$ of the quad-rotor 2 has a drift of -0.015), and the fault $\bar{m}_2 = m_{12} = 0.01$ is injected in the quad-rotor 1 at $t = 150$ seconds ($\tilde{m}_\phi$ of the quad-rotor 1 has a drift of 0.01). As shown in Figure 4.4, the diagnoser state first changes to s_D at $t = 58$ seconds after the occurrence of the wind gust with no false alarms generated. As depicted in Figure 4.3, once the vehicles are moving through the wind gust area, all the residuals expect r_7 will pass the threshold and based on Table 4.6, the diagnoser detects the occurrence of the wind gust. Later on when a fault is injected in the first actuator of the quad-rotor 2, the diagnoser state switches to s_3 at 123.2 seconds based on the coding set Ω^f_3. Finally, after the occurrence of the fault in the second actuator of the quad-rotor 1 at $t = 150$ seconds, the diagnoser switches to the state $s_{2,3}$ at $t = 163$ seconds based on the coding set $\Omega^f_3 \cap \Omega^f_2$ (note that τ_0 is selected as 10 seconds.)

Consequently, we can conclude that the HFD scheme can perfectly detect and isolate concurrent faults despite the presence of large concurrent disturbances. However, if one *only* uses the first six residuals $r_1, ..., r_6$ according to [31, 120, 73], with the threshold values $J^2_{th_k}$, although no false alarms will be generated due to large wind gust but the faults at the quad-rotors 1 and 2 *cannot* be detected and isolated. However, by using the proposed hybrid FDI methodology, one is able to distinguish the occurrence of concurrent faults as well as large disturbances by designing only one additional residual signal.

Figure 4.5 depicts the residual evaluation functions associated with occurrence of a fault in the first actuator of the quad-rotor 3 with $\bar{m}_5 = m_{31} = 0.01$ at $t = 70$ seconds while the quad-rotors are passing through the area between $x = 50$ and $x = 100$ meters and are subject to the wind gust. Figure 4.6 shows the states corresponding to the HFD scheme. As seen from this figure, the diagnoser first detects the occurrence of a large disturbance between $t = 58.6$ seconds and $t = 71.8$ seconds and its state is changed to s_D, then it detects the occurrence of a fault at $t = 71.8$ seconds and changes its state to $s_{F,D}$ based on the residual r_7. At $t = 88.2$ seconds it detects the removal of the wind gust and changes its state to s_F, and finally at $t = 126.3$ seconds it can isolate the fault and its state is changed to s_5 based on the coding set Ω_5^f. It should be noted that the occurrence of the wind gust is also detected by the diagnoser while its state is at s_F when the diagnoser switches between s_F and $s_{F,D}$ states during $t = 88.2$ seconds and $t = 126.3$ seconds.

The above fault scenario demonstrates the capability of the proposed FDI algorithm for detecting the occurrence of a fault while the quad-rotors are subjected to large disturbances and no false alarms are generated due to the presence of disturbances. Moreover, it is demonstrated that if one uses conventional approaches in the literature that are based on only the first six residuals, faults *cannot* be detected and isolated in the network of unmanned vehicles consisting of multiple quad-rotors. Finally, it should be emphasized that all the above simulations are conducted by using the full nonlinear dynamics of the quad-rotors while the linear residual generators are only incorporated in the proposed hybrid fault diagnoser. This also highlights that the proposed approach is robust with respect to uncertainties and unmodeled dynamics arising from the linearization approximation process.

4.4 Hybrid FDI Design for the Almost-Lighter-Than-Air-Vehicle (ALTAV) System

The Example 4.1 and the network of unmanned vehicles that are worked out in Sections 4.2 and 4.3 belong to a class of systems with not strongly detectable family of fault signatures. In this section, we consider the application of the proposed hybrid FDI methodology to an ALTAV system where the actuator fault signatures are strongly detectable.

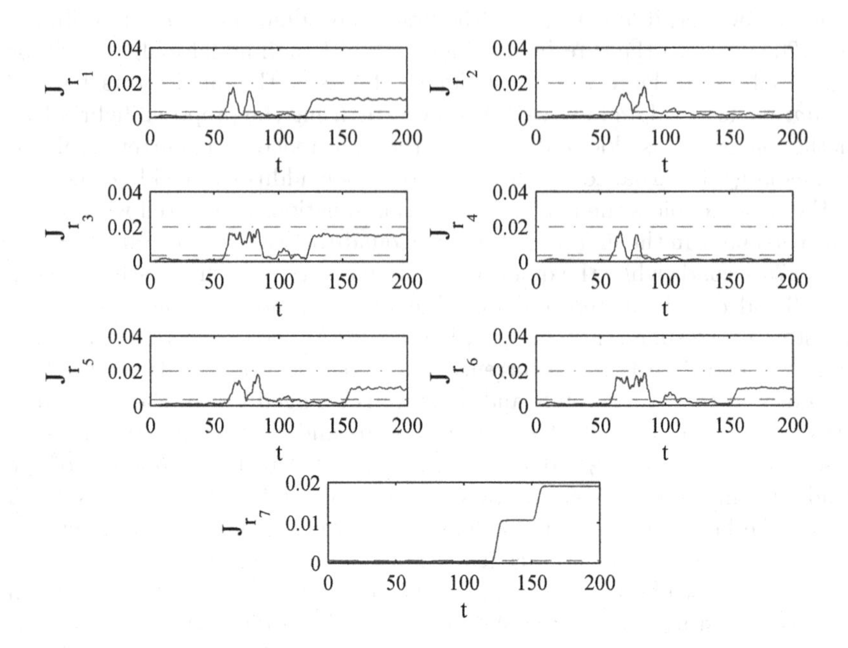

Fig. 4.3 Residual evaluation functions corresponding to concurrent faults in the first actuator of the quad-rotor 1 and the second actuator of the quad-rotor 2 (©IEEE 2010), reprinted with permission.

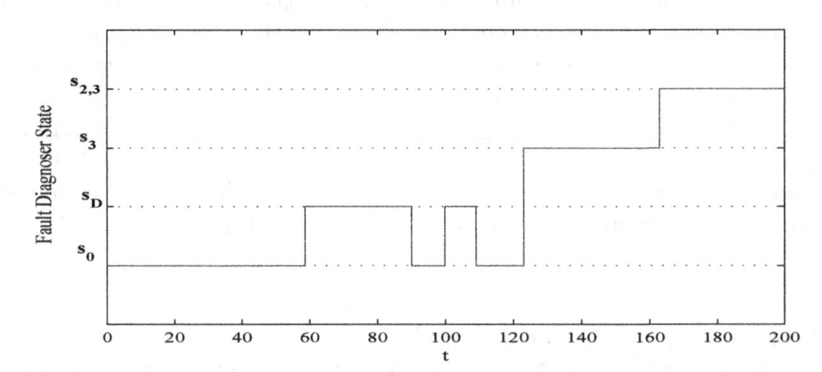

Fig. 4.4 Fault diagnoser states corresponding to concurrent faults in the first actuator of the quad-rotor 1 and the second actuator of the quad-rotor 2 (©IEEE 2010), reprinted with permission.

4.4.1 The ALTAV System

The ALTAV system considered in this section is a six degrees of freedom unmanned aerial vehicle. The states/variables describing the motion of the

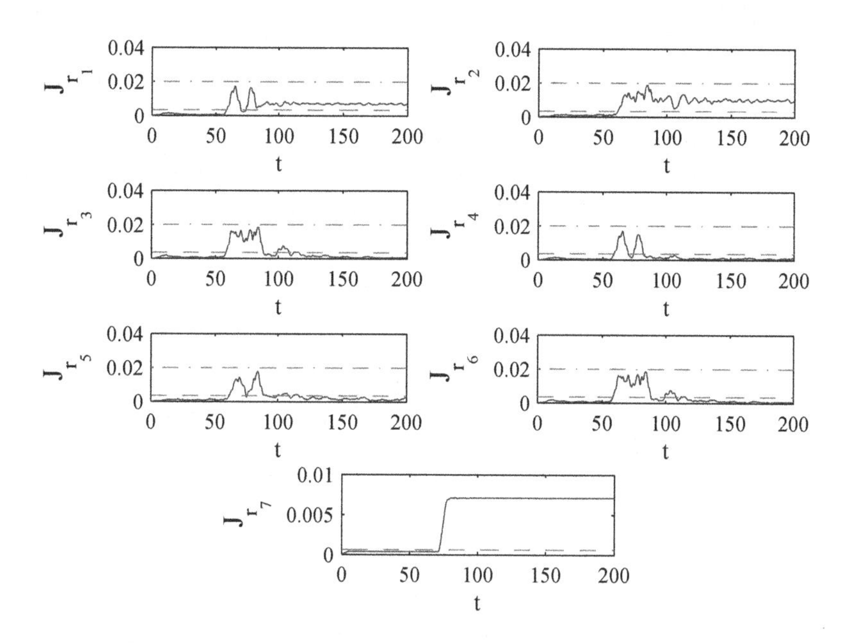

Fig. 4.5 Residual evaluation functions corresponding to a fault in the first actuator of the quad-rotor 3 (©IEEE 2010), reprinted with permission.

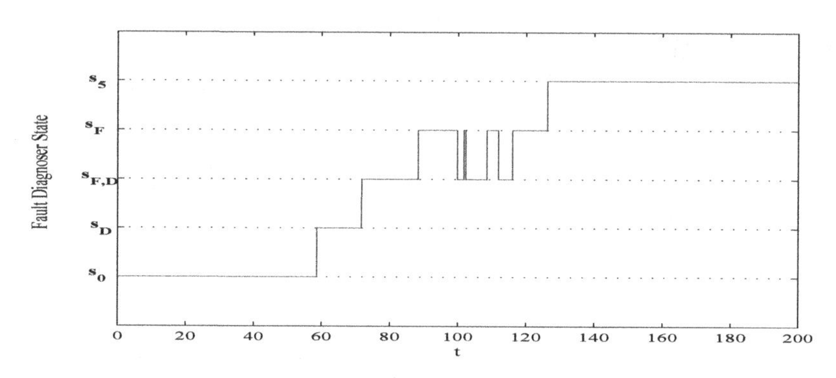

Fig. 4.6 Fault diagnoser states corresponding to a fault in the first actuator of the quad-rotor 3 (©IEEE 2010), reprinted with permission.

system are x, y, z, θ, γ and ϕ. These states correspond to the translation in the x, y and z directions and rotations about the z, y and x axes (heading, pitch and roll angles) in the local horizontal/local vertical frame, respectively. It is assumed that these states and their first order derivatives are available for measurement. It should be pointed out that the system uses a "right handed" coordinate system with the positive z direction as down. The dynamics of the ALTAV system is governed by the following equations [60]

$$\begin{aligned}
M\ddot{x} &= \sum_{i=1}^{4} F_i sin(\gamma) - C_x\dot{x} + W_x \\
M\ddot{y} &= \sum_{i=1}^{4} F_i sin(\phi) - C_y\dot{y} + W_y \\
M\ddot{z} &= -\sum_{i=1}^{4} F_i cos(\gamma)cos(\phi) - F_B + Mg - C_z\dot{z} \\
J_\theta\ddot{\theta} &= (F_1 l - F_2 l + F_3 l - F_4 l)sin(\rho) - C_\theta\dot{\theta} \\
J_\gamma\ddot{\gamma} &= (F_1 l - F_3 l) - F_B L_B sin(\gamma) - C_\gamma\dot{\gamma} \\
J_\phi\ddot{\phi} &= -(F_2 l - F_4 l) - F_B L_B sin(\phi) - C_\phi\dot{\phi}
\end{aligned} \tag{4.29}$$

where the physical significance and definition of the states and parameters are provided in Table 4.8.

Table 4.8 The ALTAV system parameters

M	Mass, kg	θ	Heading angle, Radian
J	Moment of inertia, kg m^2	γ	Pitch angle, Radian
x	Translation in x direction, m	ϕ	Roll angle, Radian
y	Translation in y direction, m	W	Wind disturbance, N
z	Translation in z direction, m	F_B	Buoyant force, N
C	Drag coefficient	F_i	Force of propeller, N
l	Perpendicular distance between the motors and vehicle center of gravity, m	ρ	Angular offset from vertical of the motor thrust vectors

The ALTAV system has four input forces $F_i, i = 1, ..., 4$ that are produced by propellers which are controlled through four vectoring brushless DC motors subject to the constraints $0 \leq F_i \leq F_i^{max}$.

4.4.2 Design of a Hybrid FDI Scheme for the ALTAV System

In this section, a bank of residual generators is first designed for the four input channels of the ALTAV system. The state space representation of the

ALTAV system is rewritten as follows

$$\begin{aligned} \dot{X} &= f(X) + \sum_{i=1}^{4} g_i(X)F_i + \sum_{j=1}^{3} p_j(X)\omega_j \\ Y &= X + v \end{aligned} \tag{4.30}$$

where $F_1, ..., F_4$ are the input force control channels, $X^\top = [x, y, z, \dot{x}, \dot{y}, \dot{z}, \theta, \gamma, \phi, \dot{\theta}, \dot{\gamma}, \dot{\phi}]$, ω_1 and ω_2 represent the wind disturbances in the x and the y directions, respectively, ω_3 represents a change in the buoyant force F_B, v is the measurement noise and

$$p_1(X) = \left[0\,0\,0\,\tfrac{1}{M}\,0\,0\,0\,0\,0\,0\,0\,0\right]^\top \tag{4.31}$$

$$p_2(X) = \left[0\,0\,0\,0\,\tfrac{1}{M}\,0\,0\,0\,0\,0\,0\,0\right]^\top \tag{4.32}$$

$$p_2(X) = \left[0\,0\,0\,0\,0\,\tfrac{1}{M}\,0\,0\,0\,0\,-\tfrac{L_B sin(X_8)}{J_\gamma}\,-\tfrac{L_B sin(X_9)}{J_\phi}\right]^\top \tag{4.33}$$

where we denote $X = \{X_i\}_{i=1}^{12}$ and $Y = \{Y_i\}_{i=1}^{12}$.

First, we need to generate the residual signals $r_i, i = 1, ..., 4$ such that each residual signal r_i is only affected by F_i and is decoupled from all other faults $F_j, j \neq i$. Towards this end, the largest observability codistributions $\Pi_i^* = o.c.a.((\sum_*^{\mathcal{L}_i})^\perp)$ should be found where $\mathcal{L}_1 = span\{g_2(X), g_3(X), g_4(X)\}$, $\mathcal{L}_2 = span\{g_1(X), g_3(X), g_4(X)\}$, $\mathcal{L}_3 = span\{g_1(X), g_2(X), g_4(X)\}$, and $\mathcal{L}_4 = span\{g_1(X), g_2(X), g_3(X)\}$ such that

$$span\{g_i(X)\} \not\subset (\Pi_i^*)^\perp, \quad i = 1, ..., 4 \tag{4.34}$$

Condition (4.34) is the necessary condition for decoupling faults in a nonlinear system. For the ALTAV system since full state measurements $X_i, i = 1, ..., 12$ are assumed to be available, we have $\Pi_i^* = \mathcal{L}_i^\perp$.

According to the above observability codistribution, the following set of states can be found such that $z_i, i = 1, ..., 4$ is affected by F_i and is decoupled from the other input channels $F_j, j \neq i$, namely

$$\begin{aligned} z_1 &= 2J_\gamma sin(\rho)sin(X_8)X_{11} + J_\theta sin(X_8)X_{10} + lM sin(\rho)X_4 \\ z_2 &= -2J_\phi sin(\rho)sin(X_8)X_{12} - J_\theta sin(X_8)X_{10} + lM sin(\rho)X_4 \\ z_3 &= -2J_\gamma sin(\rho)sin(X_8)X_{11} + J_\theta sin(X_8)X_{10} + lM sin(\rho)X_4 \\ z_4 &= 2J_\phi sin(\rho)sin(X_8)X_{12} - J_\theta sin(X_8)X_{10} + lM sin(\rho)X_4 \end{aligned}$$

The state space representation of the ALTAV system corresponding to the above states is now expressed as follows

$$
\begin{aligned}
\dot{z}_1 =& -2C_\gamma sin(\rho)sin(X_8)X_{11} - 2F_B L_B sin(\rho)sin^2(X_8) - C_\theta X_{10} sin(X_8) \\
&+ 2J_\gamma cos(X_8)sin(\rho)X_{11}^2 + J_\theta X_{10}X_{11}cos(X_8) - C_x X_4 l sin(\rho) \\
&+ 4l sin(\rho)sin(X_8)F_1 + l sin(\rho)\omega_1 - 2L_B sin(\rho)sin^2(X_8)\omega_3 \\
\dot{z}_2 =& 2C_\phi sin(\rho)sin(X_8)X_{12} + 2F_B L_B sin(\rho)sin(X_8)sin(X_9) \\
&- 2J_\phi cos(X_8)sin(\rho)X_{11}X_{12} + C_\theta X_{10} sin(X_8) - J_\theta X_{10}X_{11}cos(X_8) \\
&- C_x X_4 l sin(\rho) + 4l sin(\rho)sin(X_8)F_2 + l sin(\rho)\omega_1 \\
&+ 2L_B sin(\rho)sin(X_8)sin(X_9)\omega_3 \\
\dot{z}_3 =& 2C_\gamma sin(\rho)sin(X_8)X_{11} + 2F_B L_B sin(\rho)sin^2(X_8) - 2J_\gamma cos(X_8)sin(\rho)X_{11}^2 \\
&- C_\theta X_{10} sin(X_8) + J_\theta X_{10}X_{11}cos(X_8) - C_x X_4 l sin(\rho) \\
&+ 4l sin(\rho)sin(X_8)F_3 + l sin(\rho)\omega_1 + 2L_B sin(\rho)sin^2(X_8)\omega_3 \\
\dot{z}_4 =& -2C_\phi sin(\rho)sin(X_8)X_{12} - 2F_B L_B sin(\rho)sin(X_8)sin(X_9) \\
&+ 2J_\phi cos(X_8)sin(\rho)X_{11}X_{12} + C_\theta X_{10} sin(X_8) - J_\theta X_{10}X_{11}cos(X_8) \\
&- C_x X_4 l sin(\rho) + 4l sin(\rho)sin(X_8)F_4 \\
&+ l sin(\rho)\omega_1 - 2L_B sin(\rho)sin(X_8)sin(X_9)\omega_3
\end{aligned}
$$

From the above equations, we have $\Lambda_i = \{p_1, p_3\}, i = 1, ..., 4$, and according to Remark 4.2, one needs only to generate one extra residual signal that is decoupled from ω_1 and ω_3. Towards this end, the largest observability codistribution $\Pi_5^* = o.c.a.((\sum_*^{\mathcal{L}_5})^\perp)$ should be found where $\mathcal{L}_5 = \{p_1, p_3\}$ such that

$$span\{g_i(X)\} \not\subset (\Pi_5^*)^\perp, \quad i = 1, ..., 4$$

For the ALTAV model, we have

$$\Pi_5^* = \mathcal{L}_5^\perp$$

Based on Π_5^* the following set of states can be found that are decoupled from p_1 and p_3 and are affected by all the control inputs F_i's, namely

$$
\begin{aligned}
z_5 &= X_{10} \\
z_6 &= X_5
\end{aligned}
$$

However, only one of the above states is sufficient and in order to satisfy Assumption 4.3, we can only select z_5 for this purpose, since $\Omega_3^p = \{r_6\}$ where r_6 corresponds to the residual signal that is generated by the observer of state z_6. The state space representation and the governing equation corresponding to z_5 is given by

$$\dot{z}_5 = -\frac{C_\theta}{J_\theta} z_5 - \frac{l sin(\rho)}{J_\theta}(F_1 - F_2 + F_3 - F_4)$$

In the next step, detection filters or nonlinear observers are designed for the complete set of the five states z_1 to z_5. Given the original assumption regarding the availability of all the ALTAV states, the following observers are now constructed

$$\begin{aligned}
\dot{\hat{z}}_1 =& -2C_\gamma sin(\rho)sin(Y_8)Y_{11} - 2F_BL_Bsin(\rho)sin^2(Y_8) + 2J_\gamma cos(Y_8)sin(\rho)Y_{11}^2 \\
& - C_\theta Y_{10}sin(Y_8) + J_\theta Y_{10}Y_{11}cos(Y_8) - C_xY_4lsin(\rho) \\
& + 4lsin(\rho)sin(Y_8)F_1 + k_1(z_1 - \hat{z}_1) \\
\dot{\hat{z}}_2 =& 2C_\phi sin(\rho)sin(Y_8)Y_{12} + 2F_BL_Bsin(\rho)sin(Y_8)sin(Y_9) \\
& - 2J_\phi cos(Y_8)sin(\rho)Y_{11}Y_{12} + C_\theta Y_{10}sin(Y_8) - J_\theta Y_{10}Y_{11}cos(Y_8) \\
& - C_xY_4lsin(\rho) + 4lsin(\rho)sin(Y_8)F_2 + k_2(z_2 - \hat{z}_2) \\
\dot{\hat{z}}_3 =& 2C_\gamma sin(\rho)sin(Y_8)Y_{11} + 2F_BL_Bsin(\rho)sin^2(Y_8)] - 2J_\gamma 7cos(Y_8)sin(\rho)Y_{11}^2 \\
& - C_\theta Y_{10}sin(Y_8) + J_\theta Y_{10}Y_{11}cos(Y_8) - C_xY_4lsin(\rho) \\
& + 4lsin(\rho)sin(Y_8)F_3 + k_3(z_3 - \hat{z}_3) \\
\dot{\hat{z}}_4 =& -2C_\phi sin(\rho)sin(Y_8)Y_{12} - 2F_BL_Bsin(\rho)sin(Y_8)sin(Y_9) \\
& + 2J_\phi cos(Y_8)sin(\rho)Y_{11}Y_{12} + C_\theta Y_{10}sin(Y_8) - J_\theta Y_{10}Y_{11}cos(Y_8) \\
& - C_xY_4lsin(\rho) + 4lsin(\rho)sin(Y_8)F_4 + k_4(z_4 - \hat{z}_4) \\
\dot{\hat{z}}_5 =& -\frac{C_\theta}{J_\theta}\hat{z}_5 - \frac{lsin(\rho)}{J_\theta}(F_1 - F_2 + F_3 - F_4) + k_5(z_5 - \hat{z}_5)
\end{aligned}$$

where $k_i > 0, i = 1, ..., 5$ are the observer gains that can be selected to achieve a tradeoff between ensuring higher robustness to uncertainties and disturbances versus higher sensitivity to faults. By utilizing the above observers, the residual signals $r_i(t) = z_i(t) - \hat{z}_i(t), i = 1, ..., 5$ can now be produced. The coding sets for the fault channels $F_1, ..., F_4$ and the disturbance inputs ω_1 and ω_3 are as follows: $\Omega_1^f = \{1,5\}$, $\Omega_2^f = \{2,5\}$, $\Omega_3^f = \{3,5\}$, $\Omega_4^f = \{4,5\}$ and $\Omega_1^p = \Omega_3^p = \{1,2,3,4\}$. Moreover, it is clear that the sufficient condition in Lemma 4.3 is also satisfied for the ALTAV system.

Once the residual signals $r_i(t), i = 1, ..., 5$ are constructed, the next step is to determine the threshold J_{th_i} and the evaluation function $J_{r_i}(t)$. In this section, the following evaluation functions are selected

$$J_{r_i}(t) = \int_{t-T_0}^{t} r_i^\top(t)r_i(t)dt, \quad i = 1, ..., 5 \tag{4.35}$$

where T_0 is the length of the evaluation window. The main advantage of these evaluation functions is that one can detect intermittent faults easily. The threshold values and the corresponding residual logic units $R_i^1, i = 1, ..., 5$, and $R_i^2, i = 1, ..., 4$, are selected according to equations (4.15), (4.17), (4.18) and (4.20), respectively.

Remark 4.5. The evaluation function $J_{r_i}(t) = r_i(t)$ <u>is not</u> applicable to the ALTAV system since oscillations are present in the residual signals $r_i, i = 1, ..., 4$, when simultaneously there is a fault in one of the input channels F_i and the state γ happens to be also varying about zero. Under these circumstances the corresponding residual signals will also behave similar to that of γ. Refer to Figure 4.10 that is depicted in the next section for further clarification.

The next step is to design a DES fault diagnoser H. According to Table 4.1, the state set is defined as $S_H = \{s_0, ..., s_4, s_{1,2}, ..., s_{3,4}, s_{1,D}, ..., s_{4,D}, s_F, s_{F,D}\}$, where the cardinality of S is 23. The input set of the diagnoser is

$$I = \{R_1^1, ..., R_5^1, R_1^2, ...R_4^2\}$$

and the output set is equal to S_H. The transition function λ_H can be found following the results in Section 4.4.2, which is shown in Table 4.9.

Table 4.9 Transition functions of the ALTAV DES fault diagnoser

Current State	Input R_1^1 ... R_5^1 R_1^2 ... R_4^2	Next state
s_0	$R_1^1 \bigwedge R_5^1 = 1$	s_1
s_0	$R_2^1 \bigwedge R_5^1 = 1$	s_2
s_0	$R_3^1 \bigwedge R_5^1 = 1$	s_3
s_0	$R_4^1 \bigwedge R_5^1 = 1$	s_4
s_0	$R_1^1 \bigwedge R_2^1 \bigwedge R_3^1 \bigwedge R_4^1 = 1$	s_D
s_0	$R_5^1 = 1$	s_F
s_1	all zero	s_0
s_1	$R_1^1 \bigwedge R_2^1 \bigwedge R_5^1 = 1$	$s_{1,2}$
s_1	$R_1^1 \bigwedge R_3^1 \bigwedge R_5^1 = 1$	$s_{1,3}$
s_1	$R_1^1 \bigwedge R_4^1 \bigwedge R_5^1 = 1$	$s_{1,4}$
s_1	$\bigwedge_{j=1}^5 R_j^1 = 1$	$s_{1,D}$
⋮	⋮	⋮
s_4	all zero	s_0
s_4	$R_1^1 \bigwedge R_4^1 \bigwedge R_5^1 = 1$	$s_{1,4}$
s_4	$R_2^1 \bigwedge R_4^1 \bigwedge R_5^1 = 1$	$s_{2,4}$
s_4	$R_3^1 \bigwedge R_4^1 \bigwedge R_5^1 = 1$	$s_{3,4}$
s_4	$\bigwedge_{j=1}^5 R_j^1 = 1$	$s_{4,D}$
s_D	all zero	s_0
s_D	$R_1^2 \bigwedge R_2^1 \bigwedge R_3^1 \bigwedge R_4^1 \bigwedge R_5^1 = 1$	$s_{1,D}$
s_D	$R_2^2 \bigwedge R_1^1 \bigwedge R_3^1 \bigwedge R_4^1 \bigwedge R_5^1 = 1$	$s_{2,D}$
s_D	$R_3^2 \bigwedge R_1^1 \bigwedge R_2^1 \bigwedge R_4^1 \bigwedge R_5^1 = 1$	$s_{3,D}$
s_D	$R_4^2 \bigwedge R_1^1 \bigwedge R_2^1 \bigwedge R_3^1 \bigwedge R_5^1 = 1$	$s_{4,D}$
s_D	$R_1^1 \bigwedge R_2^1 \bigwedge R_3^1 \bigwedge R_4^1 \bigwedge R_5^1 = 1$	$s_{F,D}$
s_F	all zero	s_0
s_F	$R_1^1 \bigwedge R_5^1 = 1$	s_1
s_F	$R_2^1 \bigwedge R_5^1 = 1$	s_2
s_F	$R_3^1 \bigwedge R_5^1 = 1$	s_3
s_F	$R_4^1 \bigwedge R_5^1 = 1$	s_4
s_F	$R_1^1 \bigwedge R_2^1 \bigwedge R_3^1 \bigwedge R_4^1 \bigwedge R_5^1 = 1$	$s_{F,D}$

4.4.3 Simulation Results

In this section, simulation results of the proposed hybrid scheme that is applied to the nonlinear ALTAV system are presented. Various actuator faults are considered in the four input channels of the ALTAV system. Figure 4.7 shows the desired reference trajectory and the actual ALTAV trajectory in the normal or healthy operation of the system. In this surveillance-type maneuvering mission, the ALTAV initiates its motion from the coordinate (0,0,10) (m), follows a rectangular path in the x-y plane, and then changes its altitude to 20 (m) and follows the same pattern in this altitude. Tables 4.10 and

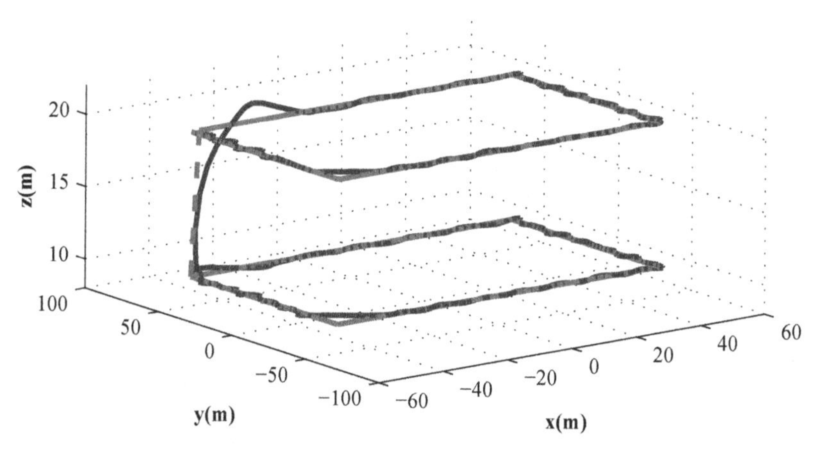

Fig. 4.7 The reference trajectory (dashed line) and the ALTAV trajectory (solid line) corresponding to a healthy operation.

4.11 show the characteristics of the output measurement noise and tolerable disturbance inputs $\mathfrak{D}_1$ that are considered in the simulation studies below, respectively. Since large changes in the buoyant force F_B do not produce any changes in the residuals, only the wind disturbance in the x-direction needs to be considered for simulations (the wind disturbance in the y-direction is unobservable from the residual signals), i.e.,

$$\mathfrak{D}_2 = \{W_x | 0.5 \ll W_x < 5(Newton)\}$$

By considering the worst case scenario of the residuals corresponding to the healthy mode of the ALTAV system subject to the measurement noise and tolerable input disturbances $\mathfrak{D}_1$ (refer to Table 4.11), the threshold values of $J^1_{th_i} = 6e-4, J^2_{th_i} = 4.5e-3,\ i = 1, ..., 4$ and $J^1_{th_5} = 3e-5$ and the evaluation window $T_0 = 5$ seconds and $T_0 = 10$ seconds are selected for the residual signals $r_1, ..., r_4$ and r_5, respectively. It should be pointed out that since the residual signal r_5 is decoupled from all the disturbance inputs, one can select a lower threshold value for it.

Table 4.10 The output measurement noise characteristics

Output measurement	Noise characteristics
γ	Uniform random variable $\pm$ 1 degree
ϕ	Uniform random variable $\pm$ 1 degree
$\dot{x}$	Uniform random variable $\pm$ 0.1 m/s
$\dot{\theta}$	Uniform random variable $\pm$ 0.5 degree/s
$\dot{\gamma}$	Uniform random variable $\pm$ 0.5 degree/s
$\dot{\phi}$	Uniform random variable $\pm$ 0.5 degree/s

Table 4.11 The tolerable disturbance input $\mathfrak{D}_1$ characteristics

	Disturbance characteristics
W_x	Uniform random variable $\pm$ 0.5 Newton
W_y	Uniform random variable $\pm$ 0.5 Newton
ΔF_B	Uniform random variable $\pm$ 2 Newton

Figure 4.8 shows the residual evaluation functions corresponding to a permanent float fault in the input channel F_1 at $t = 100$ seconds. A concurrent wind disturbance gust that is represented by a rectangular pulse of a constant amplitude 3 (N) in the x-direction between $t = 80$ and $t = 120$ seconds is also applied to the ALTAV system in this simulation. Figure 4.9 depicts the state of the DES fault diagnoser. As shown in this figure, the diagnoser state first changes to s_D at $t = 83$ seconds after the occurrence of a large wind disturbance in the x-direction with no false alarms generated. Later when a fault in the input channel F_1 is injected, the diagnoser state first switches to $s_{F,D}$ at $t = 101.3$ seconds and then after about 2 seconds it switches to the state $s_{1,D}$ at $t = 103$ seconds. Consequently, one can conclude that the diagnoser can perfectly detect and isolate the fault despite the presence of a large concurrent disturbance. Finally, after the disturbance is removed at $t = 120$ seconds, the diagnoser switches to the state s_1. It should be emphasized that if one *only* uses the first 4 residuals $r_1, ..., r_4$, not only a *false alarm* will be generated due to the wind disturbance but also the actual fault at $t = 100$ seconds *cannot* be detected and isolated. However, by using the proposed hybrid FDI methodology, we are able to distinguish the occurrence of a fault as well as large wind disturbance in the x-direction ($\omega_1 \in \mathfrak{D}_2$) by designing only one additional residual signal.

Figure 4.10 shows the residual signals corresponding to the above fault scenario. According to the points raised in Remark 4.5, it is now evident from Figure 4.10 why one cannot directly evaluate the residual signals by using their thresholds, i.e. $J_{r_i} = r_i$ since due to the dynamics of the ALTAV system after the occurrence of a fault the residuals tends to oscillate in and out of the threshold bounds.

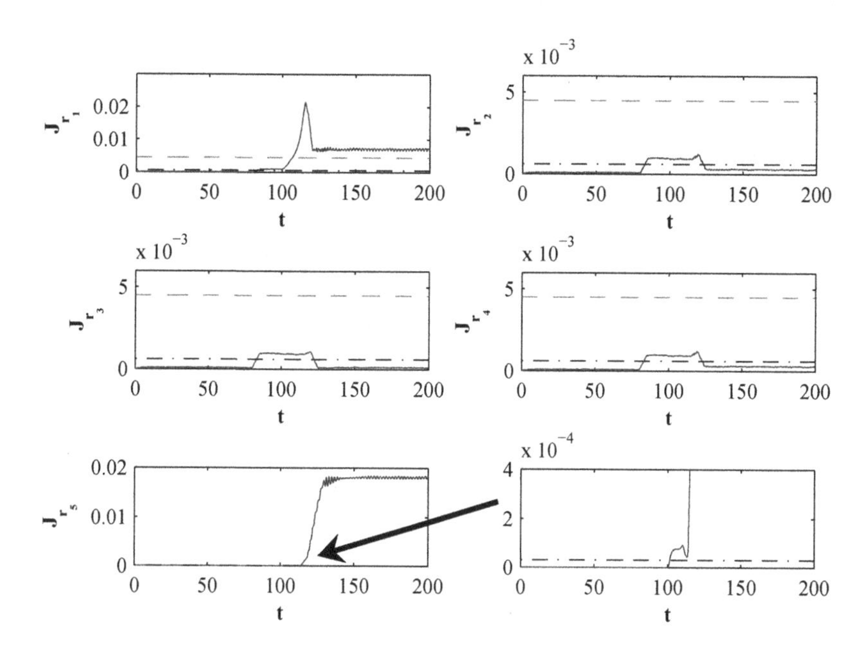

Fig. 4.8 Residual evaluation functions corresponding to a float fault in the F_1 actuator (the dashed dots correspond to the threshold values $J^1_{th_i} = 6e-4, i = 1, ..., 4$ and $J^1_{th_5} = 3e-5$, the dashed lines correspond to the threshold value $J^2_{th_i} = 4.5e-3$) [135].

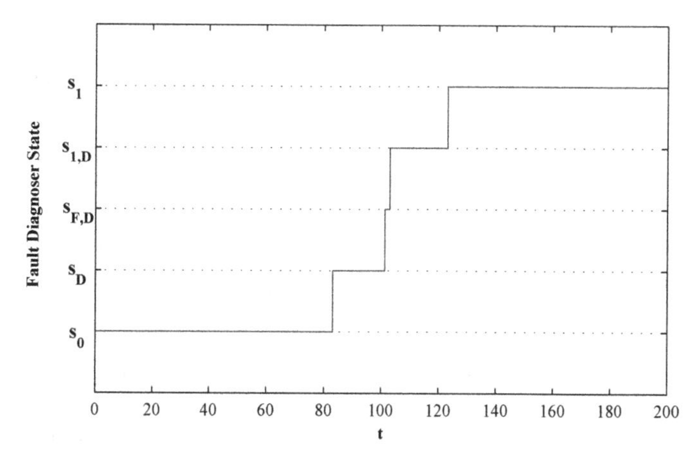

Fig. 4.9 Fault diagnoser states corresponding to a float fault in the F_1 actuator [135].

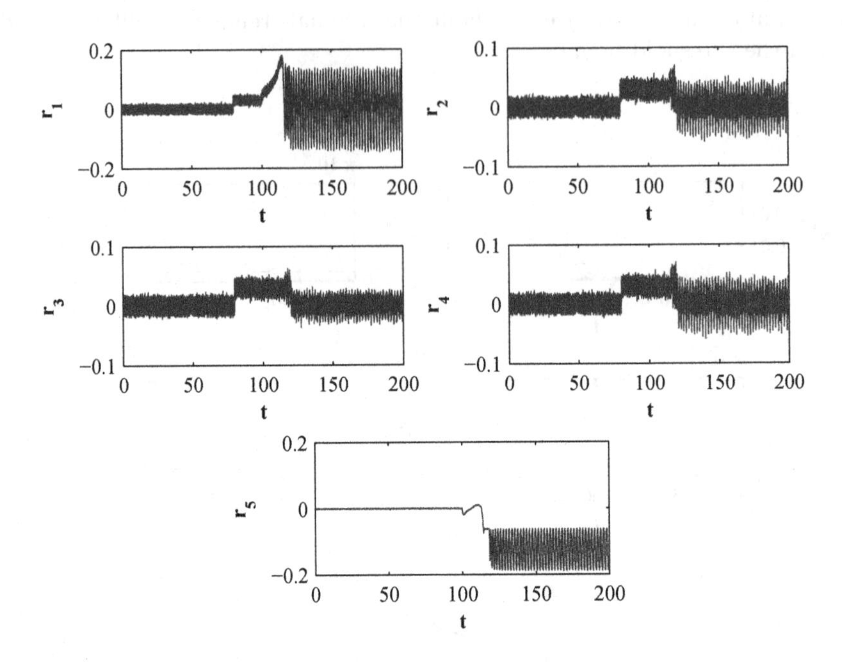

Fig. 4.10 Residuals corresponding to a float fault in the F_1 actuator.

Figure 4.11 depicts the residual evaluation functions associated with a permanent hard over fault (HOF) in the input channel F_2 that is applied at $t = 100$ seconds, and Figure 4.12 shows the corresponding fault diagnoser state. A wind disturbance gust that is represented by a rectangular pulse of constant amplitude 3 (N) in the x-direction is also injected between $t = 50$ seconds and $t = 80$ seconds in the simulations. As seen from Figure 4.12, the diagnoser first detects the fault at $t = 100.2$ seconds and also isolates the fault at $t = 107.5$ seconds. Moreover, no false alarms are generated due to the presence of the disturbance input. As shown in Figure 4.11, the residual evaluation function J_{r_2} does not exceed the second threshold values $J^2_{th_i}$. Consequently, if one *only* uses a conventional FDI method and chooses the threshold values by considering the entire disturbance set, i.e. $\mathfrak{D}_2$, then the hard over fault *cannot* be detected and isolated. However, by using the proposed approach, this fault can easily be detected and then isolated despite the presence of large wind disturbances.

Figures 4.13 and 4.14 depict the residual evaluation functions and the fault diagnoser state, respectively, corresponding to a permanent 50% loss of effectiveness (LOE) fault in the input channel F_3 injected at $t = 100$ seconds and a wind gust disturbance that is represented by a rectangular pulse of constant amplitude 3 (N) in the x-direction between $t = 130$ seconds and $t = 160$ seconds. As seen from the above figures, the residual evaluation

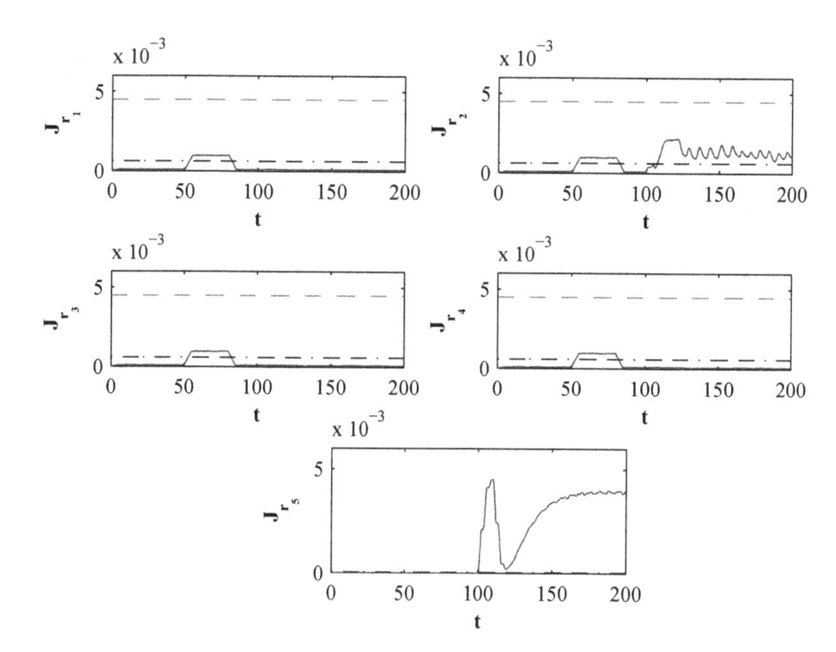

Fig. 4.11 Residual evaluation functions corresponding to a hard over fault in the F_2 actuator (the dashed dots correspond to the threshold values $J^1_{th_i} = 6e-4, i = 1, ..., 4$ and $J^1_{th_5} = 3e-5$, the dashed lines correspond to the threshold value $J^2_{th_i} = 4.5e-3$) [135].

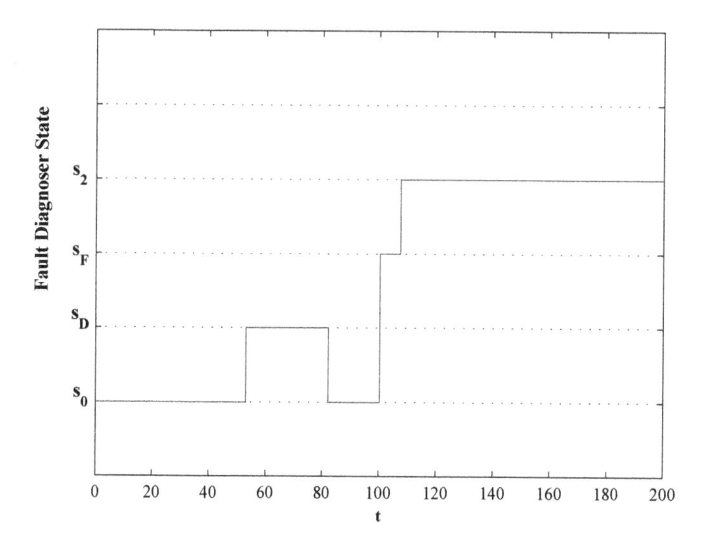

Fig. 4.12 Fault diagnoser states corresponding to a hard over fault in the F_2 actuator [135].

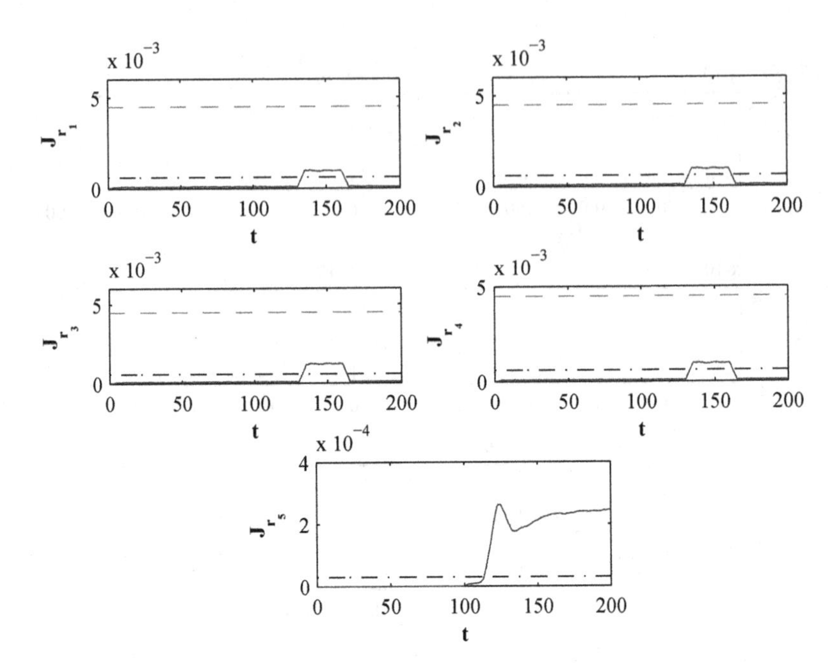

Fig. 4.13 Residual evaluation functions corresponding to a 50% loss of effectiveness in the F_3 actuator (the dashed dots correspond to the threshold values $J^1_{th_i} = 6e - 4, i = 1, ..., 4$ and $J^1_{th_5} = 3e - 5$, the dashed lines correspond to the threshold value $J^2_{th_i} = 4.5e - 3$).

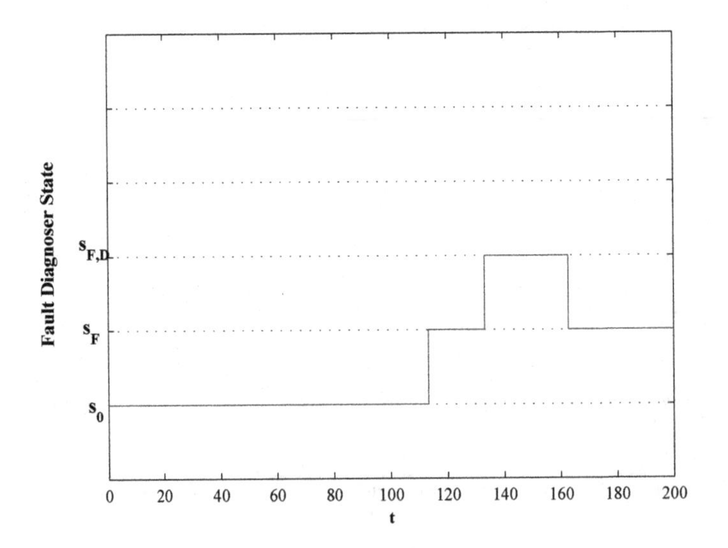

Fig. 4.14 Fault diagnoser states corresponding to a 50% loss of effectiveness in the F_3 actuator.

function J_{r_3} does not exceed its threshold whereas the residual evaluation function J_{r_5} does exceed its threshold. In this scenario the fault diagnoser can only detect the occurrence of the fault but cannot isolate it. Moreover, the occurrence of a concurrent large disturbance does not cause any false alarms in any of the input channels. This scenario also illustrates another advantage of introducing the set of complementary residuals (the residual r_5 in the case of the ATLAV system), since by incorporating this set of residuals one can also detect the occurrence of a low severity fault.

The last scenario we consider corresponds to presence of multiple faults in the ALTAV actuators. For the generated residual signals, we have $\Omega_i^f \cup \Omega_j^f \subset \Omega_i^f \cup \Omega_1^p$. Therefore, according to Remark 4.4, a waiting time interval of $\tau_0 = 1$ second is considered for detecting the second fault. Figures 4.15 and 4.16 show the corresponding residual evaluation functions and the fault diagnoser, respectively, to an intermittent float fault in the input channel F_1 that is applied between $t = 100$ seconds and $t = 150$ seconds, a permanent 50% loss of effectiveness fault in the input channel F_4 that is applied at $t = 120$ seconds and a wind gust disturbance that is represented by a rectangular pulse of constant amplitude 3 (N) that is injected in the x-direction between $t = 50$ seconds and $t = 80$ seconds. According to these figures one does clearly detect and isolate multiple faults in the ALTAV system.

4.5 Conclusions

A novel hybrid fault detection and isolation scheme is proposed for both linear and nonlinear systems subject to large disturbances. The proposed scheme consists of two modules, namely, a bank of residual generators and a discrete-event system (DES)-based fault diagnoser. A DES diagnoser is developed which uses the residual signals and their temporal behavior to robustly detect and isolate the faulty channels. The proposed hybrid FDI methodology is applied to the problem of actuators fault detection and isolation for an ALTAV system and a network of unmanned vehicles. Simulation results clearly illustrate and demonstrate the effectiveness and advantages of the developed framework and methodology.

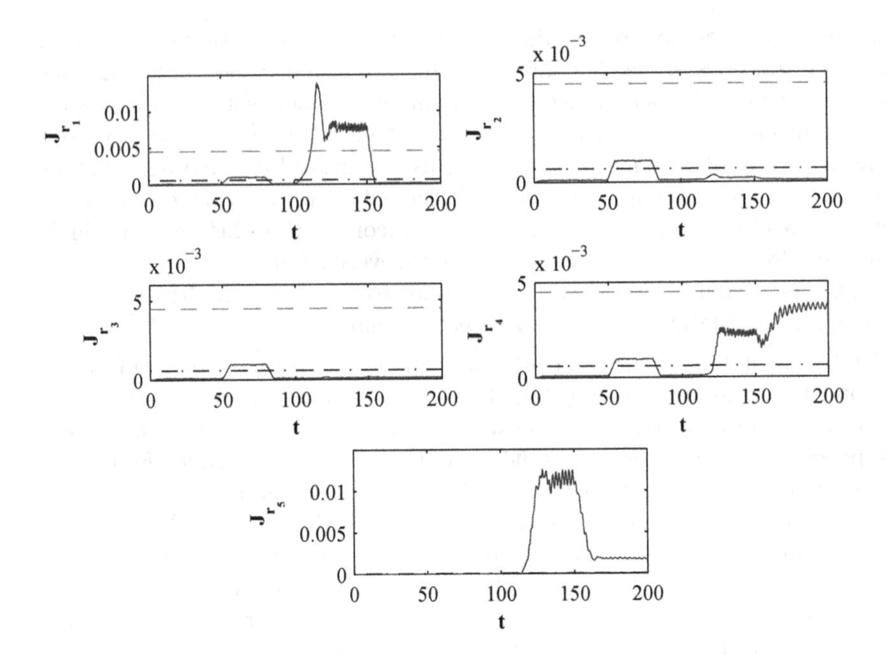

Fig. 4.15 Residual evaluation functions corresponding to multiple faults in the F_1 and F_4 actuators (the dashed dot corresponds to the threshold values $J^1_{th_i} = 6e-4, i = 1, ..., 4$ and $J^1_{th_5} = 3e-5$, the dashed lines correspond to the threshold value $J^2_{th_i} = 4.5e-3$) [135].

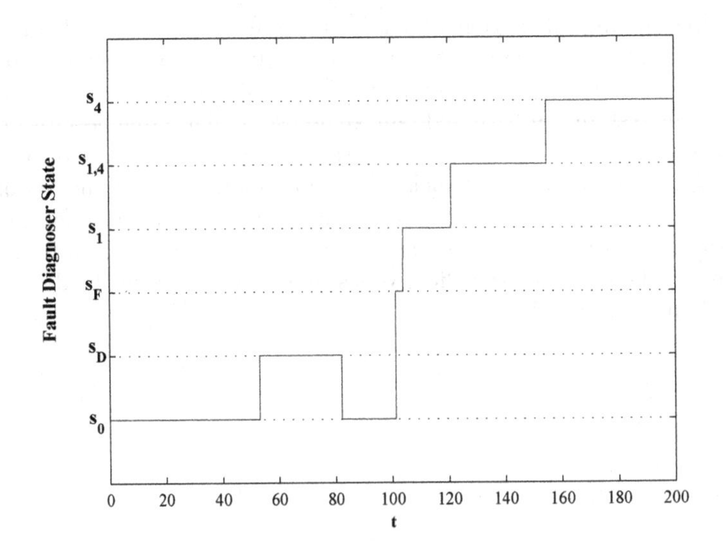

Fig. 4.16 Fault diagnoser states corresponding to multiple faults in the F_1 and F_4 actuators [135].

Chapter 5
Compensating for Communication Channels Effects in the FDI Problem

This chapter deals with the problem of fault detection and isolation in a network of unmanned vehicles when there exist imperfect communication channels among the vehicles. A discrete-time communication link having a stochastic packet dropping effect is considered based on the Gilbert-Elliott model [62, 78] which is known as the packet erasure channel model. It is shown that the entire network can be modeled as a discrete-time Markovian Jump System (MJS). This problem is then treated in the general framework of Markovian jump systems. This motivates us to develop a geometric FDI framework for both continuous-time and discrete-time Markovian jump systems and apply the developed scheme to FDI of a network of unmanned vehicles in presence of imperfect communication channels. The work presented in this chapter has partially appeared in [130, 137, 133, 134, 132].

This chapter is organized as follows. After a brief overview of the relevant literature, we begin in Section 5.2 by discussing the packet erasure channel model that is used in this chapter. In Section 5.3, a discrete-time Markovian jump model of a network of unmanned vehicles integrated with the packet erasure channel model is developed corresponding to the centralized architecture. In Section 5.4, the FDI problem for discrete-time MJS is investigated in a geometric FDI framework. Toward this end, in Section 5.4.1 a geometric property related to the unobservable subspace of a Markovian jump system is presented. A new approach for determining the conditions for weak-observability of MJS systems is then introduced. The concept of unobservability subspace is presented and an algorithm for obtaining it is described. In Section 5.4.2, the necessary and sufficient conditions for solvability of the fundamental problem of residual generation (FPRG) for MJS systems are obtained by utilizing the introduced unobservability subspace. We also present and develop sufficient conditions for designing H_∞-based FDI algorithms for MJS systems subject to input disturbances in Section 5.4.3. The proposed algorithm is then applied to the FDI problem in a formation flight of satellites. In Section 5.5 similar results are derived for FDI of continuous-time Markovian jump systems. Finally, the proposed algorithm is applied to fault detection and isolation of the VTOL (vertical take-off and landing) helicopter actuators in Section 5.5.4.

5.1 Introduction

Recently, networked control systems (NCS) have become a hot area of research and have found some important applications in a wide variety of engineering systems including manufacturing plants, aircraft, automobiles, etc. Generally speaking, NCS are comprised of a large number of actuators, sensors, and controllers that are all equipped with network interfaces. All the signals, including real-time sensing, controller outputs, commands, coordination and supervision signals are transmitted through shared network channels. Novel and interesting challenges arise when feedback loops are closed within this networked architecture. The network itself is a dynamical system that exhibits characteristics such as networked-induced delay, packet dropout, asynchronous clock among network nodes that could all degrade the performance of the closed-loop system and even destabilize the system.

Research on NCS has received increasing attention in recent years, and a large body of work has been produced on the modeling, design and stability analysis. However, few results exist on the fault detection and diagnosis in NCS. In [206], a Kalman filter is designed for fault detection with the assumption that network delay is known but time-varying. In [205], fault detection for NCS with missing measurements is investigated. In this work NCS is modeled as a Markovian jump linear system and an H_∞ optimization technique is used for designing a fault detection filter for the Markovian jump system. In [203], a robust parity space approach is developed for *fault detection* in NCS with unknown and time-varying delays between the controller and the actuator. In [3], NCS model is transformed into a framework of linear time-invariant systems with modeling uncertainty that is caused by stochastic changes in the system parameters due to the network induced delay and data loss. A unified approach that is proposed in [55] is used for residual generation. In [109], the effects of network induced delay are modeled as time-varying disturbances and the fault isolation filter for the NCS with multiple faults is parameterized based on directional residual generation approach. The remaining degrees of freedom are then used to satisfy an H_∞ disturbance attenuation property in the framework of Markovian jump systems. In [115], the FDI problem for NCS with large transfer delays is considered. By employing the multirate sampling method together with the augmented state matrix method, the NCS with large transfer delays is modeled as a Markovian jump system and an H_∞ fault detection filter is designed for the developed model.

The packet delivery characteristics of a network can be modeled as a Bernoulli or two-state Markov process [62, 78, 163, 197, 81, 104, 195, 94].

The latter is commonly used for modeling the fading communication channels and is also known as the *packet erasure channel model.* The integration of a discrete-time plant with the channel models yield a discrete-time Markovian jump system (MJS).

A great deal of attention has recently been devoted to the Markovian jump systems [168, 26, 52, 196, 24, 48] which comprise an important class of hybrid systems. This family of systems is generally modeled by a set of linear systems with transitions among models that are determined by a Markov chain taking values in a finite set. Markovian jump systems are also popular in modeling many practical systems where one may experience abrupt changes in system structure and parameters. These changes are quite common and do frequently occur in manufacturing systems, economic systems, communication systems, power systems, etc. [47].

In recent years, only a few work on fault detection and isolation of MJS systems have appeared in the literature. In [205, 209, 115], a robust fault detection (and *not an isolation*) filter for discrete-time Markovian jump systems is developed based on an H_∞ filtering framework, in which the residual generator is also an MJS system. An LMI approach is developed for solving the problem. In [184], a robust fault identification filter for a class of discrete-time Markovian jump systems with mode dependent time-delays and norm bounded uncertainty is developed based on an H_∞ optimization technique. In the approach in [184], the generated residual signal is an estimate of the fault signal. However, the problem of fault isolation of Markovian jump systems has not been completely solved and fully addressed in the above references.

In this chapter, based on the Gilbert-Elliott model [62, 78, 163, 197], a network of unmanned vehicles integrated with a two-state Markov process model of communication channels is modeled as a discrete-time Markovian jump system. The FDI problem in a network of unmanned vehicles in the presence of imperfect communication links can be solved in the framework of Markovian jump systems. A geometric approach is adopted for the FDI problem of both discrete-time and continuous-time Markovian jump systems.

Towards this end, in this chapter we first derive a geometric property for the unobservable subspace of MJS systems (Theorems 5.2 and 5.9) and develop a new approach for determining its weak-observability (Algorithm 5.1). The notion of an unobservability subspace is then introduced for MJS systems (Definition 5.6). To construct an algorithm for obtaining this subspace, an alternative definition of an unobservability subspace is presented in Theorem 5.4, which only depends on the matrices of the system. Based on this alternative definition, an algorithm for constructing the smallest unobservability subspace containing a given subspace is proposed (Algorithm 5.3). By utilizing the developed geometric framework, necessary and sufficient solvability conditions are derived for Markovian jump systems (Theorem 5.6). By using the properties of unobservability subspaces, a set of residuals is then generated such that each residual is affected by one fault and is decoupled from others. Finally, the problem of designing an H_∞-based FDI algorithm for a

MJS system that has an unknown transition matrix and is subjected to external disturbances is investigated. These results are obtained by applying the proposed geometric approach and the H_∞ disturbance attenuation technique (Theorem 5.7).

5.2 The Packet Erasure Channel Model

In this chapter, discrete-time communication links with stochastic packet dropping effects are considered among the vehicles and the FDI unit. Many unique phenomena are introduced in a networked system due to the presence of communication links such as [81]

1. Time delay: Data is usually buffered, quantized and coded before transmission over a communication channel. Following a network induced delay due to propagation, the data is received and decoded in the receiver end. In cases where the data is not received properly, the communication protocol may re-transmit the data. Hence, by the time the information is used by the receiver, a delay has been introduced. This delay is usually random with a time-varying probability distribution.
2. Data loss: In most communication protocols, if the data is not received within a specific time limit, the packet is assumed to be lost. A variety of factors and reasons can cause this data loss. For instance, simultaneous transmission by two transmitters in a shared multiple access medium such as wireless channels, may lead to loss of both transmitter data. In a network of communication channels, overflow of buffers can also cause the packet loss. Finally, in a network of unmanned vehicles, the presence of obstacles among vehicles and limited line of sight of each vehicle may lead to packet losses.
3. Quantization: In digital communication networks any data needs to be quantized and the number of bits that can be transmitted at every sampling time is usually upper-bounded.
4. Data corruption: Due to noise and attenuation introduced by the channel, the received data may not be identical to the signal that the transmitter sent through the communication channel. Most communication protocols has error detection and error correction algorithms to reduce the data corruption of the communication links.

At every sampling time k, the time line for the operation of a communication link of the i-th vehicle can be considered as follows:

1. The i-the vehicle creates a packet containing the information that are needed to be shared with other vehicles or the FDI unit.
2. The packet is sent across the link.
3. At time step $k+1$ the packet is either received without error or dropped stochastically.

In the above communication links, it is assumed that:

- The channel coding is ignored and it is assumed that the packet will be either received and decoded successfully at the end of the link or totally lost.
- If a packet arrives too late, it is discarded and treated as a dropped packet and the lost packets are not re-sent.
- The number of bits in each packet is relatively large, and hence the quantization effects of the channel are ignored.

One of the following models [94] can be considered for modeling the stochastic packet dropout:

- The independent and identically distributed (i.i.d.) Bernoulli model : A Bernoulli random variable γ_k is assigned to the communication link such that if the packet k is received correctly, then $\gamma_k = 1$, otherwise $\gamma_k = 0$. A random variable γ_k is i.i.d. with probability distribution $P(\gamma_k = 1) = \lambda$ and $P(\gamma_k = 0) = (1 - \lambda)$.
- The Gilbert-Elliott model [62, 78] as shown in Figure 5.1. In this model, the network is considered as a discrete-time Markov chain with two possible states: "good" and "bad". The packet is received correctly in the "good" state and is dropped in the "bad" state. The network jumps between these two states according to a Markov chain with transition probability matrix Π as

$$\begin{bmatrix} \pi_{00} & \pi_{01} \\ \pi_{10} & \pi_{11} \end{bmatrix}$$

where 1 is the good state, 0 is the bad state, and π_{ij} is the probability from the previous state i to the next state j.

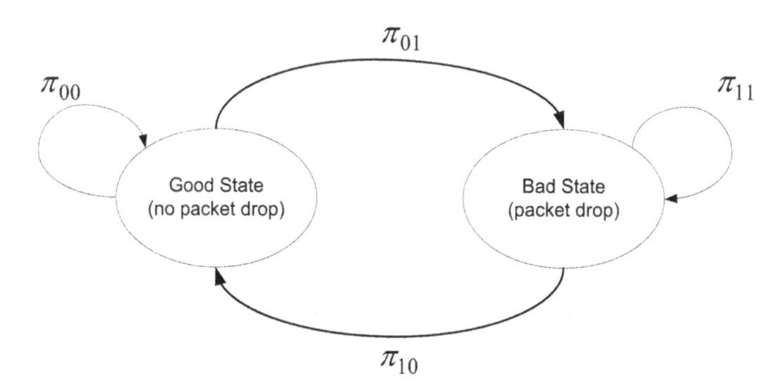

Fig. 5.1 The Gilbert-Elliott model.

In this chapter, we consider the Gilbert-Elliott model of the communication link since this model is capable of capturing the dependence between consecutive losses, i.e. bursty packet dropping.

5.3 A Network of Unmanned Vehicles: Imperfect Communication Channels

In this section, a network of unmanned vehicles is considered with the packet erasure channel model that is represented as a discrete-time MJS. It should be noted that a centralized architecture is considered here. Due to the discrete feature of the communication links, the discrete model of each vehicle is considered as follows

$$\begin{aligned} x_i(k+1) &= Ax_i(k) + Bu_i(k) + \sum_{j=1}^{a} L_j m_{ij}(k) \\ z_{ij}(k) &= C(x_i(k) - x_j(k)) \qquad j \in \mathrm{N}_i \end{aligned} \tag{5.1}$$

having the sampling period T_s, where the set $\mathrm{N}_i \subset \mathbf{N} \setminus i$ represents the set of vehicles that vehicle i can sense and is designated as the neighboring set of vehicle i, and $z_{ij} \in \mathcal{Z}_i, j \in \mathrm{N}_i$ represents the state measurement relative to the other vehicles. Let $\mathrm{N}_i = \{i_1, i_2, ..., i_{|\mathrm{N}_i|}\}$. We have $z_i(t) = [z_{ii_1}^\top(t), z_{ii_2}^\top(t), \cdots, z_{ii_{|\mathrm{N}_i|}}^\top(t)]^\top$, and equation (5.1) can be rewritten as $z_i(t) = C_i x(t)$, where $x(t) = [x_1^\top(t), x_2^\top(t), \cdots, x_N^\top(t)]^\top$.

As discussed in Section 3.3.1, in the centralized architecture, all the information should be sent to a central FDI unit through the "entire" communication network channels. It is assumed that each vehicle sends its information i.e. u_i and z_i with a sampling period T_s. Therefore, the overall system can be modeled as follows

$$\begin{aligned} x(k+1) &= A^N x(k) + B^N u(k) + \sum_{k=1}^{N}\sum_{j=1}^{a} \bar{L}_{kj} m_{kj}(k) \\ z(k) &= \bar{C} x(k) \end{aligned} \tag{5.2}$$

where $A_N = I_N \otimes A$, $B_N = I_N \otimes B$, $\otimes$ denotes the Kronecker product, N is a positive integer, I_N is an $N \times N$ identity matrix, $\bar{L}_{kj}$ is the $(k-1) \times a + j$ column of B^N, and

$$u = \begin{bmatrix} u_1 \\ \vdots \\ u_N \end{bmatrix}, \quad z = \begin{bmatrix} z_1 \\ \vdots \\ z_N \end{bmatrix}, \quad \bar{C} = \begin{bmatrix} C_1 \\ \vdots \\ C_N \end{bmatrix}. \tag{5.3}$$

The central FDI unit waits for T_s seconds to collect all the information from all the vehicles. Therefore, a network delay of less than T_s does not affect the performance of the FDI algorithm. If the packet has a delay of more than T_s, then the FDI unit assumes that there is a packet dropout in that specific communication link between the central FDI unit and the vehicle.

In order to model packet dropout of the communication network channels, a stochastic variable λ_i is assigned to the channel between the i-th vehicle and the central vehicle and it represents data communication status, i.e., $\lambda_i(k) = 1$ implies that the measurement u_i and z_i at time k arrives correctly, while $\lambda_i(k) = 0$ implies that this measurement is lost. λ_i can be modeled as a discrete-time Markov chain with two state $\Psi_i = \{0, 1\}$ with the transition probability matrix $\Pi_i = (\pi_{ij})_{l,j\in\Psi_i}$, where π_{ij} is defined as

$$\pi_{lj} = Pr\{\lambda(k+1) = j|\lambda(k) = l\}$$

When the current information (u_i and z_i) of the i-th vehicle does not arrive on time ($\lambda_i(k) = 0$), the FDI unit uses the last available measurement of the control signal u_i and the available output measurements z_j's, $j \neq i$. In other words, if we denote $u_i^a(k)$ as the i-th vehicle information that is used by the central FDI unit, then

$$u_i^a(k) = \begin{cases} u_i(k) & \text{if } \lambda_i(k) = 1 \\ \text{the last available measurment} & \text{if } \lambda_i(k) = 0 \end{cases}$$

Moreover, the central FDI unit uses the available relative state measurements $z(k) = \bar{C}_i x(k)$ where

$$\bar{C}_i = \begin{bmatrix} C_1 \\ \vdots \\ C_{i-1} \\ 0 \\ C_{i+1} \\ \vdots \\ C_N \end{bmatrix} \tag{5.4}$$

Remark 5.1. It should be pointed out that the available relative state measurements for $\lambda_i(k) = 0$ is $z^i(k) = \bar{C}^i x(k)$ having a variable dimension where

$$\bar{C}^i = \begin{bmatrix} C_1 \\ \vdots \\ C_{i-1} \\ C_{i+1} \\ \vdots \\ C_N \end{bmatrix} \tag{5.5}$$

However, it can be shown that the geometric properties of a Markovian jump system with either the measurement matrix $\bar{C}_i$ or $\bar{C}^i$ are identical. Indeed, we have $\text{Ker}(\bar{C}_i) = \text{Ker}(\bar{C}^i)$. Moreover, let $\mathcal{S} =< \text{Ker}H^i\bar{C}^i|A + D^i\bar{C}^i >$, where $H^i = [H_1, ..., H_{i-1}, H_{i+1}, ..., H_N]$ and $D^i = [D_1, ..., D_{i-1}, D_{i+1}, ..., D_N]$, then $A+D^i\bar{C}^i = A+D_i\bar{C}_i$ where $\text{Ker}H^i\bar{C}^i = \text{Ker}H_i\bar{C}_i$ $H_i = [H_1, ..., H_{i-1}, 0, H_{i+1}, ..., H_N]$ and $D_i = [D_1, ..., D_{i-1}, 0, D_{i+1}, ..., D_N]$, and hence $\mathcal{S} =< \text{Ker}H_i\bar{C}_i|A+D_i\bar{C}_i >$. Therefore, for keeping the dimension of the output constant, the measurement matrix $\bar{C}_i$ is considered here.

In the next assumption, it is assumed that the change of the control input for each vehicle corresponding to each FDI sampling period is bounded.

Assumption 5.1 *It is assumed that for each vehicle the quantity* $\theta_i^u(k) = u_i^a(k) - u_i(k)$ *is an* $\mathfrak{L}_2$ *bounded signal.*

Therefore, in the case of $\lambda_i(k) = 0$, the overall system can be modeled as

$$
\begin{aligned}
x(k+1) &= A^N x(k) + B^N u(k) + \sum_{k=1}^{N}\sum_{j=1}^{a} \bar{L}_{kj} m_{kj}(k) + B_i^d \theta_i^u(k) \\
z(k) &= \bar{C}_i x(k)
\end{aligned}
\tag{5.6}
$$

where

$$
B_i^d = \begin{bmatrix} 0 \\ \vdots \\ B_i \\ \vdots \\ 0 \end{bmatrix}
$$

In the case of $\lambda_i(k) = 1$, the overall system model is given as in (5.2). Therefore, the disturbance terms $B_i^d \theta_i^u(k)$ are added due to the loss of u_i and different measurement matrix $\bar{C}_i$ is used due to the loss of z_i.

In order to incorporate the data dropout from all the communication channels, a discrete-time Markov chain λ with $N+1$ states having the transition probability matrix $\Pi = (\pi_{ij})_{i,j \in \Psi}$ is introduced. Without loss of generality, it is assumed that only one data dropout is allowed at each sampling period. This assumption only limits the number of modes of the system to $N+1$ and can be easily relaxed by increasing the number of system modes and by considering multiple data dropout. The mode 1 corresponds to the case with no packet dropout in all the communication links and the mode $i+1$ corresponds to the packet dropout in the communication links of the i-th vehicle. Therefore, the entire system can be modeled by the following Markovian jump system

$$\begin{aligned} x(k+1) &= A^N x(k) + B^N u(k) + \bar{B}^d_{\lambda(k)} \bar{\theta}^u(k) + \sum_{k=1}^{N} \sum_{j=1}^{a} \bar{L}_{kj} m_{kj}(k) \\ z(k) &= \bar{C}_{\lambda(k)} x(k) \end{aligned} \tag{5.7}$$

where

$$\bar{\theta}^u(k) = \begin{bmatrix} \theta^u_1 \\ \vdots \\ \theta^u_N \end{bmatrix}$$

and for $\lambda(k) = i + 1$, $i \in \mathbf{N}$ we have

$$\bar{B}^d_{i+1} = \begin{bmatrix} 0 & \cdots & 0 & \cdots \\ \vdots & & \vdots & \\ 0 & \cdots & B_i & \cdots \\ \vdots & & \vdots & \end{bmatrix}$$

where B_i is the i-th row and the i-th column of matrices $\bar{B}^d_{i+1}$. For $\lambda(k) = 1$, we have $\bar{B}^d_1 = 0$ and $\bar{C}_0 = \bar{C}$.

Remark 5.2. It should be noted that the FDI problem in sensor networks (Figure 5.2) can be treated similarly in the framework of Markovian jump systems. Consider a discrete-time linear system

$$x(k+1) = Ax(k) + Bu(k) + B_d d(k) \tag{5.8}$$

where $x(k) \in \mathcal{X}$ is the state, $u(k) \in \mathcal{U}$ is the control input, and $d(k)$ is the input disturbance. The state of the plant is measured by N sensors where the i-th sensor generates the output measurements according to

$$y_i(k) = C_i x(k) + D_{di} d(k) \tag{5.9}$$

Every sensor sends its own measurement through the communication channel. By combining the packet erasure channel model with each measurement equation, a discrete-time MJS model of the entire sensor network can be obtained as follows

$$\begin{aligned} x(k+1) &= Ax(k) + Bu(k) + B_d d(k) \\ y(k) &= C_{\lambda(k)} x(k) + D_{\lambda(k)} d(k) \end{aligned} \tag{5.10}$$

Consequently, the developed strategy in the next section will be applicable to both networks of unmanned vehicles and sensor networks.

In the next section, a geometric FDI framework for discrete-time Markovian jump systems is developed.

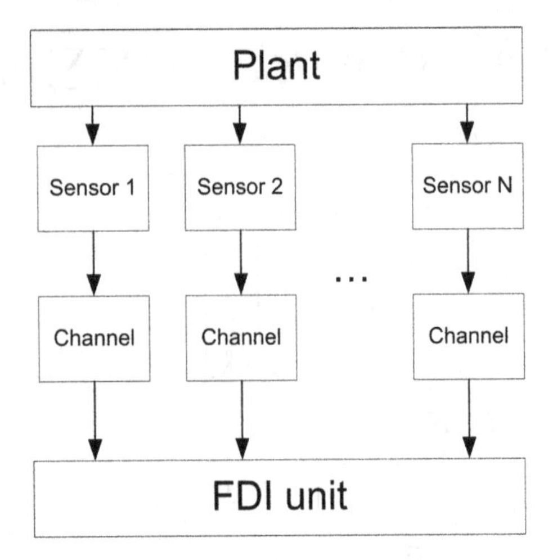

Fig. 5.2 The FDI problem in sensor networks.

5.4 Discrete-time Markovian Jump Systems (MJS)

Consider the following discrete-time Markovian jump system

$$\begin{aligned} x(k+1) &= A_{\lambda(k)}x(k) + B_{\lambda(k)}u(k) \\ y(k) &= C_{\lambda(k)}x(k), \quad x(0) = x_0, \quad \lambda(0) = i_0 \end{aligned} \tag{5.11}$$

where $x \in \mathcal{X}$ is the state of the system with dimension n; $u \in \mathcal{U}, y \in \mathcal{Y}$ are input and output signals with dimensions m and q, respectively; and $\{\lambda(k), k \geq 0\}$ is a discrete-time irreducible Markov process taking values in the finite set $\Psi = \{1, ..., N\}$. The Markov process describes the switching between the different system modes and its evolution is governed by the following probability transitions:

$$\pi_{ij} = \mathbb{P}\{\lambda(k+1) = j | \lambda(k) = i\}$$

where $\sum_{j=1}^{N} \pi_{ij} = 1$.

It is assumed that $\pi_{ii} > 0$, $i \in \Psi$. The matrices $A_{\lambda(k)}$, $B_{\lambda(k)}$ and $C_{\lambda(k)}$ are known constant matrices for all $\lambda_k = i \in \Psi$. For simplicity, we denote the matrices associated with $\lambda(k) = i$ by $A_{\lambda(k)} = A_i$, $B_{\lambda(k)} = B_i$, and $C_{\lambda(k)} = C_i$. Furthermore, the MJS system (5.11) is represented by $(\mathfrak{A}, \mathfrak{B}, \mathfrak{C}, \Pi)$, where $\mathfrak{A} = (A_1, ..., A_N)$, $\mathfrak{B} = (B_1, ..., B_N)$, $\mathfrak{C} = (C_1, ..., C_N)$ and $\Pi = [\pi_{ij}]$, $i, j \in \Psi$.

5.4.1 Unobservable and Unobservability Subspaces

In this section, a geometric definition for an unobservable subspace of discrete-time Markovian jump systems (MJS) is introduced. The notion of unobservability subspace is then formalized for discrete-time MJS systems that are governed by (5.11). To develop an algorithm for constructing this subspace, an alternative definition of unobservability subspace is presented, which only depends on the matrices of the system. Based on this alternative definition, an algorithm for constructing the smallest unobservability subspace containing a given subspace is proposed. As shown in the next section, the unobservability subspace for the MJS system plays a central role in solving the fault detection and isolation problem of Markovian jump systems.

We first start with the definition of weak observability for the Markovian jump system (5.11) with $B_i = 0, i \in \Psi$.

Definition 5.1 ([46]) *The system $(\mathfrak{A}, \mathfrak{C}, \Pi)$ is said to be weakly (W-) observable when there exists $\gamma > 0$ such that*

$$W^{n^2 N}(x_0, i) = \mathbb{E}\Big\{x_0^\top \Gamma(n^2 N) x_0 | x(0) = x_0, \lambda_0 = i_0\Big\} \geq \gamma |x_0|^2, \quad (5.12)$$

$\forall x_0 \in \mathcal{X}$, $i_0 \in \Psi$ where the point-wise observability Grammian Γ is defined as

$$\Gamma(k) = \sum_{t=0}^{k} \Phi^\top(t) C_{\lambda(t)}^\top C_{\lambda(t)} \Phi(t) \quad (5.13)$$

and Φ is the state transition matrix of the system $(\mathfrak{A}, \mathfrak{B}, \mathfrak{C}, \Pi)$.

In [46], a collection of matrices $\mathcal{O} = \{\mathcal{O}_1, ..., \mathcal{O}_N\}$ is introduced for testing the W-observability of Markovian jump systems according to the following procedure. Let $O_i(0) = C_i^\top C_i$, $i \in \Psi$ and define the sequence of matrices as

$$O_i(k) = A_i^\top (\sum_{j=1}^{N} \pi_{ij} O_j(k-1)) A_i \quad k > 0, i \in \Psi \quad (5.14)$$

Then the matrix $\mathcal{O}_i$ is defined according to

$$\mathcal{O}_i = [O_i(0)\ O_i(1) \cdots O_i(n^2 N - 1)]^\top \quad (5.15)$$

Theorem 5.1 ([46]) *The MJS system (5.11) is W-observable if and only if $\mathcal{O}_i$ has a full rank for each $i \in \Psi$.*

Definition 5.2 *A state x is said to be unobservable if $W^{n^2 N}(x, i) = 0$ for all $i \in \Psi$.*

Let $\mathcal{Q}$ denotes the unobservable set of the MJS system (5.11), i.e.,

$$\mathcal{Q} = \{x | W^{n^2 N}(x,i) = 0, \forall i \in \Psi, t \geq 0\} \tag{5.16}$$

It is shown in [46] (Lemma 3) that for irreducible Markov processes, $\mathcal{N}\{\mathcal{O}_i\} = \mathcal{N}\{\mathcal{O}_j\}$, $i,j \in \Psi$, $i \neq j$ and $\mathcal{Q} = \mathcal{N}\{\mathcal{O}_i\}$. Therefore, $\mathcal{Q}$ is the subspace of $\mathcal{X}$ and is called the unobservable subspace of the Markovian jump system (5.11). The theorem introduced below characterizes a geometric property of $\mathcal{Q}$.

Theorem 5.2 *An unobservable subspace Q for the system $(\mathfrak{A}, \mathfrak{C}, \Pi)$ with the irreducible Markov process is the largest A_i-invariant ($i \in \Psi$) that is contained in $\mathcal{K} = \bigcap_{i=1}^{N}$ Ker C_i.*

Proof: It follows from the above discussion that $\mathcal{Q} \subseteq$ Ker C_i, $i \in \Psi$, and hence $\mathcal{Q} \subseteq \mathcal{K}$. Let $x \in \mathcal{Q}$. Our goal is to show that $A_i x \in \mathcal{Q}$ for all $i \in \Psi$ (i.e. $\mathcal{Q}$ is A_i-invariant, $i \in \Psi$). Since $x \in \mathcal{N}\{O_i(k-1)\}$ and $x \in \mathcal{N}\{O_i(k)\}$, $i \in \Psi$, then

$$\left.\begin{array}{r} x^\top O_i(k)x = x^\top A_i^\top (\sum_{j=1}^{N} \pi_{ij} O_j(k-1)) A_i x = 0 \\ \pi_{ii} > 0 \end{array}\right\} \Rightarrow O_i(k-1)A_i x = 0$$

Consequently, $A_i x \in \mathcal{N}\{O_i(k-1)\}$ for all $k > 0$ and $A_i x \in \mathcal{N}\{\mathcal{O}_i\}$ for all $i \in \Psi$. This shows that $\mathcal{Q}$ is A_i-invariant for all $i \in \Psi$.

Next, we show that Q is the largest A_i-invariant ($i \in \Psi$) that is contained in $\mathcal{K}$. Let $\mathcal{V}$ be an A_i-invariant ($i \in \Psi$) subspace that is contained in $\mathcal{K}$. Clearly, $\mathcal{V} \subseteq \mathcal{N}\{O_i(0)\}$, $i \in \Psi$. Let $\mathcal{V} \subseteq \mathcal{N}\{O_i(k-1)\}$, $i \in \Psi$ and $x \in \mathcal{V}$, then

$$x^\top O_i(k)x = A_i^\top (\sum_{j=1}^{N} \pi_{ij} O_j(k-1)) A_i x = 0$$

since $A_i x \in \mathcal{V}$ ($\mathcal{V}$ is A_i-invariant). Hence, $\mathcal{V} \subseteq \mathcal{N}\{O_i(k)\}$, $i \in \Psi$ and $\mathcal{V} \subseteq \mathcal{Q}$. This shows that $\mathcal{Q}$ contains all the subspaces that are A_i-invariant ($i \in \Psi$) and is contained in $\mathcal{K}$. ■

Corollary 5.1 *An unobservable subspace Q is contained in $<$ Ker $C_i | A_i >$, $i \in \Psi$.*

Proof: Note that $<$ Ker $C_i | A_i >$ is the largest A_i-invariant that is contained in Ker C_i. Therefore, we have $\mathcal{Q} \subseteq <$ Ker $C_i | A_i >, i \in \Psi$. ■

The above corollary proves that if each pair (A_i, C_i), $i \in \Psi$ is observable then $Q = 0$ and the MJS system (5.11) is W-observable. Based on the Theorem 5.2 above, the following algorithm provides a procedure for constructing the required subspace Q.

Algorithm 5.1 *The subspace Q coincides with and is obtained from the last term of the following sequence*

$$\mathcal{Z}_0 = \mathcal{K}; \quad \mathcal{Z}_i = \mathcal{K} \cap (\bigcap_{k=0}^{N} A_k^{-1} \mathcal{Z}_{i-1}), \quad (i \in \boldsymbol{k}) \tag{5.17}$$

where the value of $k \leq dim(\mathcal{K})$ *is determined from the condition* $\mathcal{Z}_{k+1} = \mathcal{Z}_k$.

Let us denote the largest A_i-invariant ($i \in \Psi$) subspace that is contained in $\mathcal{K}$ by

$$<< \mathcal{K} | A_i >>_{i \in \Psi} \tag{5.18}$$

Consider the system $(\mathfrak{A}, \mathfrak{C}, \mathfrak{B}, \Pi)$ that is governed by (5.11). Let $\mathcal{Q}$ be A_i-invariant $i \in \Psi$ and $\mathcal{Q} \subseteq$ Ker C_i, and $P : \mathcal{X} \to \mathcal{X}/\mathcal{Q}$ be a canonical projection. The factor system is defined by $(\mathfrak{A} : \mathcal{X}/\mathcal{Q}, \mathfrak{B} : \mathcal{X}/\mathcal{Q}, \mathfrak{C} : \mathcal{X}/\mathcal{Q}, \Pi)$ with $\mathfrak{A} : \mathcal{X}/\mathcal{Q} = (A_1 : \mathcal{X}/\mathcal{Q}, ..., A_N : \mathcal{X}/\mathcal{Q}), \mathfrak{B} : \mathcal{X}/\mathcal{Q} = (B_1 : \mathcal{X}/\mathcal{Q}, ..., B_N : \mathcal{X}/\mathcal{Q})$ $\mathfrak{C} : \mathcal{X}/\mathcal{Q} = (C_1 : \mathcal{X}/\mathcal{Q}, ..., C_N : \mathcal{X}/\mathcal{Q})$, where $PA_i = (A_i : \mathcal{X}/\mathcal{Q})P$, $(B_i : \mathcal{X}/\mathcal{Q}) = PB_i, (C_i : \mathcal{X}/\mathcal{Q})P = C_i$, $i \in \Psi$. Therefore, if $\mathcal{Q}$ is the unobservable subspace of the system $(\mathfrak{A}, \mathfrak{C}, \mathfrak{B}, \Pi)$, then $(\mathfrak{A} : \mathcal{X}/\mathcal{Q}, \mathfrak{B} : \mathcal{X}/\mathcal{Q}, \mathfrak{C} : \mathcal{X}/\mathcal{Q}, \Pi)$ is observable since the unobservable subspace is factored out.

Definition 5.3 ([46]) *The system* (5.11) *is mean square (MS) stable if for each* $x_0 \in \mathcal{X}$ *and* $i_0 \in \Psi$,

$$\lim_{k \to \infty} \mathbb{E}\{||x(k)||^2\} = 0 \tag{5.19}$$

Definition 5.4 ([46]) *We say that* $(\mathfrak{A}, \mathfrak{C}, \Pi)$ *is MS-detectable when there exists* $G = \{G_1, ..., G_N\}$ *of appropriate dimension for which* $(\mathfrak{A}_G, \Pi)$ *is MS-stable, where* $\mathfrak{A}_G = \{A_1 + G_1 C_1, ..., A_N + G_N C_N\}$.

The following computational linear matrix inequalities can be used for testing the MS-detectability of a discrete-time MJS system [46]. In other words, the MS-detectability of $(\mathfrak{A}, \mathfrak{C}, \Pi)$ is equivalent to the feasibility of the set

$$\begin{bmatrix} A_i^\top X_i A_i + C_i^\top S_i^\top A_i + A^\top S_i C_i & C_i^\top S_i^\top \\ S_i C_i & -X_i \end{bmatrix} < 0 \tag{5.20}$$

$$\sum_{j=1}^{N} \pi_{ij} H_j \leq X_i \tag{5.21}$$

in the unknowns $X_i > 0$, H_i and S_i with appropriate dimensions.

We are now in a position to introduce the notions of conditioned invariant and unobservability subspaces for Markovian jump systems.

Definition 5.5 *A subspace* $\mathcal{W} \subseteq \mathcal{X}$ *is said to be conditioned invariant for the system* $(\mathfrak{A}, \mathfrak{C}, \Pi)$ *if*

$$A_i(\mathcal{W} \cap Ker\ C_i) \subseteq \mathcal{W}, \quad i \in \Psi \tag{5.22}$$

It is clear that if $\mathcal{W}$ is conditioned invariant for system $(\mathfrak{A}, \mathfrak{C}, \Pi)$, then $\mathcal{W}$ is (C_i, A_i)-invariant for all $i \in \Psi$. Therefore, there exist maps $D_i : \mathcal{Y} \to \mathcal{X}$ such that $(A_i + D_iC_i)\mathcal{W} \subseteq \mathcal{W}, \ i \in \Psi$.

We denote the class of conditioned invariant subspaces of $\mathcal{X}$ for the system $(\mathfrak{A}, \mathfrak{C}, \Pi)$ by $\mathfrak{W}(\mathfrak{A}, \mathfrak{C}, \mathcal{X})$. If $\mathcal{W} \in \mathfrak{W}(\mathfrak{A}, \mathfrak{C}, \mathcal{X})$, we write $\mathfrak{D}(\mathcal{W})$ for the class of maps $D_i : \mathcal{Y} \to \mathcal{X}$ where $(A_i + D_iC_i)\mathcal{W} \subseteq \mathcal{W}, \ i \in \Psi$. The notion of conditioned invariant subspace for system $(\mathfrak{A}, \mathfrak{C}, \Pi)$ is a dual to that of the robust maximal controlled invariant which is introduced in [9]. By duality it can be shown that $\mathfrak{W}(\mathfrak{A}, \mathfrak{C}, \mathcal{X})$ is closed under the operation of subspace intersection, and hence for any given subspace $\mathcal{L} \subseteq \mathcal{X}$, the family of conditioned invariant subspaces that contains $\mathcal{L}$ (denoted by $\mathfrak{W}(\mathfrak{A}, \mathfrak{C}, \mathcal{L})$ or simply $\mathfrak{W}(\mathcal{L})$) has an infimal element which is denoted by $\mathcal{W}^* = \inf \mathfrak{W}(\mathfrak{A}, \mathfrak{C}, \mathcal{L})$. The following algorithm can now be used for constructing $\mathcal{W}^*$.

Algorithm 5.2 *The subspace $\mathcal{W}^*$ coincides with and is obtained from the last term of the following sequence*

$$\mathcal{W}_0 = \mathcal{L}; \qquad \mathcal{W}_k = \mathcal{L} + \sum_{i=1}^{N} A_i(\mathcal{W}_{k-1} \cap Ker\ C_i)$$

Definition 5.6 *A subspace $\mathcal{S}$ is an unobservability subspace for system* (5.11) *if there exist output injection maps $D_i : \mathcal{Y} \to \mathcal{X}$ and measurement mixing maps $H_i : \mathcal{Y} \to \mathcal{Y}$ such that $\mathcal{S}$ is an unobservable subspace of system $(\tilde{\mathfrak{A}}, \tilde{\mathfrak{C}}, \Pi)$, where $\tilde{\mathfrak{A}} = \{A_1 + D_1C_1, ..., A_N + D_NC_N\}$ and $\tilde{\mathfrak{C}} = \{H_1C_1, ..., H_NC_N\}$.*

We denote the class of all unobservability subspaces in $\mathcal{X}$ for the system $(\mathfrak{A}, \mathfrak{C}, \Pi)$ by $\mathfrak{S}(\mathfrak{A}, \mathfrak{C}, \mathcal{X})$. In the following, our goal is to derive an alternative characterization for the unobservability subspace which is independent of the maps D_i and H_i as used in Definition 5.6 (the idea is similar to that in Theorem 2.1). As shown subsequently, this alternative definition provides us with means to obtain the unobservability subspaces more readily. The following lemma presents a result that is necessary for formulating our alternative definition.

Lemma 5.1 *Let $\hat{\mathcal{S}}_j \subset \mathcal{X}$ such that $KerC_j \subseteq \hat{\mathcal{S}}_j, j \in \Psi$ and $<< \bigcap_{j=1}^N \hat{\mathcal{S}}_j | A_i >>_{i \in \Psi} = \mathcal{S}$ (refer to equation* (5.18)*), then*

$$<< \bigcap_{j=1}^{N} (\mathcal{S} + Ker\ C_j) | A_i >>_{i \in \Psi} = \mathcal{S} \tag{5.23}$$

Conversely, if $<< \bigcap_{j=1}^N (\mathcal{S} + Ker\ C_j) | A_i >>_{i \in \Psi} = \mathcal{S}$, then there exist maps $H_j : \mathcal{Y} \to \mathcal{Y}, \ j \in \Psi$ such that

$$<< \bigcap_{j=1}^{N} Ker\ H_jC_j|A_i >>_{i\in\Psi}= \mathcal{S}$$

Proof: We have $\mathcal{S} \subseteq \bigcap_{j=1}^{N} \hat{\mathcal{S}}_j \subseteq \hat{\mathcal{S}}_j$ and Ker $C_j \subseteq \hat{\mathcal{S}}_j$, $j \in \Psi$, so that $\bigcap_{j=1}^{N}(\mathcal{S} + \text{Ker } C_j) \subseteq \bigcap_{j=1}^{N} \hat{\mathcal{S}}_j$. Consequently,

$$<< \bigcap_{j=1}^{N}(\mathcal{S} + \text{Ker } C_j)|A_i >>_{i\in\Psi}\subseteq<< \bigcap_{j=1}^{N} \hat{\mathcal{S}}_j|A_i >>_{i\in\Psi}= \mathcal{S}$$

On the other hand, $A_i\mathcal{S} \subseteq \mathcal{S}$ and

$$\bigcap_{j=1}^{N}(\mathcal{S} + \text{Ker } C_j) \supseteq \mathcal{S} + \bigcap_{j=1}^{N} \text{Ker } C_j \supseteq \mathcal{S}$$

Hence, $\mathcal{S} \subseteq<< \bigcap_{j=1}^{N}(\mathcal{S} + \text{Ker } C_j)|A_i >>_{i\in\Psi}$, and as a result

$$<< \bigcap_{j=1}^{N}(\mathcal{S} + \text{Ker } C_j)|A_i >>_{i\in\Psi}= \mathcal{S}$$

To show the converse part, let $\{c_1^i, ..., c_r^i\}$ be a basis for $\mathcal{S}$ + Ker C_i such that $\{c_{r-p_i}^i, ..., c_r^i\}$ is the basis for Ker C_i (dim(Ker C_i) $= p_i$). Therefore, $y_{ij} = C_ic_j^i$, $j = 1, ..., r - p_i - 1$ are independent. Let $\{y_{i1}, ..., y_{iq}\}$ be a basis for $\mathcal{Y}$, and define

$$\begin{aligned} H_iy_{ij} &= 0, \quad j = 1, ..., r - p_i - 1 \\ H_iy_{ij} &= y_{ij}, \quad j = r - p_i, ..., r \end{aligned}$$

Consequently, Ker C_i+$\mathcal{S}$ = Ker H_iC_i, and $\bigcap_{i=1}^{N}$ Ker C_i+$\mathcal{S}$ = $\bigcap_{i=1}^{N}$ Ker H_iC_i. ■

We are now in a position to state our next result.

Theorem 5.3 *A subspace $\mathcal{S}$ is an unobservability subspace for system* (5.11) *if and only if there exist maps $D_i : \mathcal{Y} \to \mathcal{X}$, such that*

$$\mathcal{S} =<< \bigcap_{j=1}^{N}(\mathcal{S} + Ker\ C_j)|A_i + D_iC_i >>_{i\in\Psi} \tag{5.24}$$

Proof: The proof follows readily from Lemma 5.1 by taking $\hat{\mathcal{S}}_j$ = Ker H_jC_j. ■

The above theorem eliminates the need for the maps H_i from the Definition 5.6 (this is similar to the equation (2.16) for a system with only a single mode). For a given unobservability subspace $\mathcal{S}$, the measurement mixing maps H_i's can be computed from $\mathcal{S}$ by solving the following equations, namely

$$\text{Ker } H_i C_i = \mathcal{S} + \text{Ker } C_i, \quad i \in \Psi \tag{5.25}$$

Next, we try to characterize the unobservability subspace by means of an algorithm that computes $\mathcal{S}$ without explicitly constructing $(D_1, ..., D_N) \in \mathfrak{D}(\mathcal{S})$. For an arbitrary subspace $\mathcal{S} \subset \mathcal{X}$, let us define a family

$$\mathcal{G}_{(\mathfrak{A},\mathfrak{C})} = \{\mathscr{S} : \mathscr{S} = \bigcap_{i=1}^{N} (\mathcal{S} + (A_i^{-1}\mathscr{S} \cap \text{ Ker } C_i))\} \tag{5.26}$$

Below, we first show that $\mathcal{G}_{(\mathfrak{A},\mathfrak{C})}$ has a unique maximal member.

Lemma 5.2 *There exists a unique element $\mathscr{S}^* \in \mathcal{G}_{(\mathfrak{A},\mathfrak{C})}$ such that $\mathscr{S} \subseteq \mathscr{S}^*$ for every $\mathscr{S} \in \mathcal{G}_{(\mathfrak{A},\mathfrak{C})}$.*

Proof: Define a sequence $\mathscr{S}^\mu \subset \mathcal{X}$ according to

$$\mathscr{S}^0 = \mathcal{X}; \qquad \mathscr{S}^\mu = \bigcap_{i=1}^{N} (\mathcal{S} + (A_i^{-1}\mathscr{S}^{\mu-1} \cap \text{ Ker } C_i)), \quad \mu \in \mathbf{n} \tag{5.27}$$

The sequence $\mathscr{S}^\mu$ is non-increasing since $\mathscr{S}^1 \subseteq \mathscr{S}^0$ and if $\mathscr{S}^\mu \subseteq \mathscr{S}^{\mu-1}$, then

$$\mathscr{S}^{\mu+1} = \bigcap_{i=1}^{N} (\mathcal{S} + (A_i^{-1}\mathscr{S}^{\mu} \cap \text{ Ker } C_i)) \subseteq \bigcap_{i=1}^{N} (\mathcal{S} + (A_i^{-1}\mathscr{S}^{\mu-1} \cap \text{ Ker } C_i)) = \mathscr{S}^\mu$$

Therefore, there exists $k \leq n$ such that $\mathscr{S}^\mu = \mathscr{S}^k$, and we set $\mathscr{S}^* = \mathscr{S}^k$. Clearly $\mathscr{S}^* \in \mathcal{G}_{(\mathfrak{A},\mathfrak{C})}$. Next, we show that $\mathscr{S}^*$ is the maximal element. Let $\mathscr{S} \subseteq \mathcal{G}_{(\mathfrak{A},\mathfrak{C})}$, then $\mathscr{S} \subseteq \mathscr{S}^0$ and if $\mathscr{S} \subseteq \mathscr{S}^\mu$, we have

$$\mathscr{S} = \bigcap_{i=1}^{N} (\mathcal{S} + (A_i^{-1}\mathscr{S} \cap \text{ Ker } C_i)) \subseteq \bigcap_{i=1}^{N} (\mathcal{S} + (A_i^{-1}\mathscr{S}^{\mu} \cap \text{ Ker } C_i)) = \mathscr{S}^{\mu+1}$$

Consequently, $\mathscr{S} \subseteq \mathscr{S}^\mu$ for all μ, and hence $\mathscr{S} \subseteq \mathscr{S}^*$. ■

The next lemma provides an important property of the maximal element $\mathscr{S}^*$ which will be used for introducing our suggested alternative characterization of the unobservability subspace of system $(\mathfrak{A}, \mathfrak{C}, \Pi)$.

Lemma 5.3 *Let $\mathcal{S} \in \mathfrak{W}(\mathfrak{A}, \mathfrak{C}, \mathcal{X})$ and $(D_1, ..., D_N) \in \mathfrak{D}$, then $\mathscr{S}^*$ is the largest $(A_i + D_iC_i)$-invariant $(i \in \Psi)$ that is contained in $\bigcap_{j=1}^{N}(\mathcal{S} + \text{ker } C_j)$, i.e., $\mathscr{S}^* = << \bigcap_{j=1}^{N}(\mathcal{S} + \text{Ker } C_j) | A_i + D_iC_i >>_{i\in\Psi}$.*

Proof: First we show that any $\mathscr{S} \in \mathcal{G}_{(\mathfrak{A},\mathfrak{C})}$ is $(A_i + D_iC_i)$-invariant $(i \in \Psi)$. We have, $\mathscr{S} = \bigcap_{i=1}^{N}(\mathcal{S} + (A_i^{-1}\mathscr{S} \cap \text{ Ker } C_i))$ and

$$\begin{aligned}(A_j + D_jC_j)\mathscr{S} &= (A_j + D_jC_j)\bigcap_{i=1}^{N}(\mathcal{S} + (A_i^{-1}\mathscr{S} \cap \text{ Ker } C_i)) \\ &\subseteq (A_j + D_jC_j)(\mathcal{S} + A_j^{-1}\mathscr{S} \cap \text{ Ker } C_j) \\ &\subseteq (A_j + D_jC_j)\mathcal{S} + A_j(A_j^{-1}\mathscr{S} \cap \text{ Ker } C_j) \\ &\subseteq \mathcal{S} + \mathscr{S} \subseteq \mathscr{S}\end{aligned}$$

where we used the relationship $\mathscr{S} = \bigcap_{i=1}^{N}(\mathcal{S} + (A_i^{-1}\mathscr{S} \cap \text{ Ker } C_i)) \supseteq \mathcal{S} + \bigcap_{i=1}^{N}(A_i^{-1}\mathscr{S} \cap \text{ Ker } C_i) \supset \mathcal{S}$ and $A_j(A_j^{-1}\mathscr{S}) \subseteq \mathscr{S}$. Therefore, $(A_j + D_jC_j)\mathscr{S} \subseteq \mathscr{S}$, $j \in \Psi$; and hence $\mathscr{S} \in \mathfrak{W}(\mathfrak{A}, \mathfrak{C}, \mathcal{X})$ and $(D_1, ..., D_N) \in \mathfrak{D}(\mathscr{S})$.

Consequently, we have $\mathscr{S}^* \in \mathfrak{W}(\mathfrak{A}, \mathfrak{C}, \mathcal{X})$. Next, we show that for any subspace $\mathcal{W}$ such that it is $(A_i + D_iC_i)$-invariant $(i \in \Psi)$ and is contained in $\bigcap_{j=1}^{N}(\mathcal{S} + \text{ker } C_j)$, we have $\mathcal{W} \subseteq \mathscr{S}^*$. If $\mathcal{W} \subseteq \mathcal{S}$, then it follows that $\mathcal{W} \subseteq \mathscr{S}^*$, since $\mathcal{S} \subseteq \mathscr{S}^*$. Therefore, we consider the case where $\mathcal{S} \subseteq \mathcal{W}$. We have

$$A_i^{-1}\mathcal{W} \cap \text{ Ker } C_i = (A_i + D_iC_i)^{-1}\mathcal{W} \cap \text{ Ker } C_i$$

and therefore

$$\bigcap_{i=1}^{N}(\mathcal{S} + A_i^{-1}\mathcal{W} \cap \text{ Ker } C_i) = \bigcap_{i=1}^{N}(\mathcal{S} + (A_i + D_iC_i)^{-1}\mathcal{W} \cap \text{ Ker } C_i)$$

It follows that $\mathcal{W} \subseteq \mathscr{S}^0$. If $\mathcal{W} \subseteq \mathscr{S}^{\mu-1}$, then

$$\begin{aligned}\mathscr{S}^{\mu} &= \bigcap_{i=1}^{N}(\mathcal{S} + (A_i^{-1}\mathscr{S}^{\mu-1} \cap \text{ Ker } C_i)) \supset \bigcap_{i=1}^{N}(\mathcal{S} + (A_i^{-1}\mathcal{W} \cap \text{ Ker } C_i)) \\ &= \bigcap_{i=1}^{N}(\mathcal{S} + (A_i + D_iC_i)^{-1}\mathcal{W} \cap \text{ Ker } C_i)) \\ &\supset \bigcap_{i=1}^{N}(\mathcal{S} + (\mathcal{W} \cap \text{Ker } C_i)) = \sum_{i=1}^{N}(\mathcal{W} \cap (\mathcal{S} + \text{Ker } C_i)) = \mathcal{W}\end{aligned}$$

where we used the fact that $\mathcal{W} \subseteq (A_i + D_iC_i)^{-1}\mathcal{W}$ and the modular distributive rule [192] (if $\mathcal{S} \subseteq \mathcal{W}$, then $\mathcal{S} + (\mathcal{W} \cap \text{Ker } C_i) = \mathcal{W} \cap (\mathcal{S} + \text{Ker } C_i)$). Consequently, $\mathscr{S}^{\mu} \supset \mathcal{W}$; and therefore $\mathscr{S}^* \supset \mathcal{W}$. This shows that $\mathscr{S}^*$ is the largest $(A_i + D_iC_i)$-invariant $(i \in \Psi)$ which is contained in $\bigcap_{i=1}^{N}(\mathcal{S} + \text{Ker } C_i)$. ■

We are in the position to introduce the proposed alternative characterization of an unobservability subspace for system $(\mathfrak{A}, \mathfrak{C}, \Pi)$.

Theorem 5.4 *Let $\mathcal{S} \subset \mathcal{X}$ and define $\mathcal{G}_{(\mathfrak{A},\mathfrak{C})}$ as in (5.26). Then $\mathcal{S} \in \mathfrak{S}(\mathfrak{A}, \mathfrak{C}, \Pi)$ if and only if*

$$\mathcal{S} \in \mathfrak{W}(\mathfrak{A}, \mathfrak{C}, \Pi) \quad (\mathcal{S} \textit{ is conditioned invariant}) \tag{5.28}$$

and

$$\mathcal{S} = \mathscr{S}^* \tag{5.29}$$

where $\mathscr{S}^*$ is the maximal element of $\mathcal{G}_{(\mathfrak{A},\mathfrak{C})}$.

Proof: **(If part)** If (5.28) and (5.29) hold, then according to Lemma 5.3, we have $\mathcal{S} =<< \bigcap_{j=1}^{N}(\mathcal{S}+\text{Ker } C_j)|A_i >>_{i\in\Psi}$, and hence using Theorem 5.3, we have $\mathcal{S} \in \mathfrak{S}(\mathfrak{A},\mathfrak{C},\Pi)$.

(Only if part) If $\mathcal{S} \in \mathfrak{S}(\mathfrak{A},\mathfrak{C},\Pi)$, it follows that $\mathcal{S} \in \mathfrak{W}(\mathfrak{A},\mathfrak{C},\Pi)$ and according to Lemma 5.3, we have $\mathcal{S} = \mathscr{S}^*$. ■

The above theorem provides a similar characteristic for the unobservability subspace of the Markovian jump system as the one that Theorem 2.1 provides for the unobservability subspace of a system having only a single mode. Next, we show that the class of unobservability subspace of system $(\mathfrak{A},\mathfrak{C},\Pi)$ is a semilattice with respect to inclusion and subspace addition, and hence the class of unobservability subspace which contains a given subspace has an infimal element. This property is crucial for the application of the unobservability subspace to the problem of fault detection and isolation of MJS systems.

Lemma 5.4 *The class of subspaces $\mathfrak{S}(\mathfrak{A},\mathfrak{C},\Pi)$ is closed under the operation of subspace intersection.*

Proof: The proof is based on the characterization of the unobservability subspace in Theorem 5.4. Let $\mathcal{S}_1, \mathcal{S}_2 \in \mathfrak{S}(\mathfrak{A},\mathfrak{C},\Pi)$. It follows that $\mathcal{S}_1 \cap \mathcal{S}_2 \in \mathfrak{W}(\mathfrak{A},\mathfrak{C},\Pi)$. Furthermore, $\mathcal{S}_j = \mathscr{S}_j^n, j = 1, 2$, where

$$\mathscr{S}_j^0 = \mathcal{X}; \qquad \mathscr{S}_j^\mu = \bigcap_{i=1}^{N}(\mathcal{S}_j + (A_i^{-1}\mathscr{S}^{\mu-1} \cap \text{ Ker } C_i)), \quad \mu \in \mathbf{n}$$

Define $\mathscr{S}$ according to

$$\mathscr{S}^0 = \mathcal{X}; \qquad \mathscr{S}^\mu = \bigcap_{i=1}^{N}((\mathcal{S}_1 \cap \mathcal{S}_2) + (A_i^{-1}\mathscr{S}^{\mu-1} \cap \text{ Ker } C_i)), \quad \mu \in \mathbf{n}$$

We have, $\mathscr{S}_1^0 = \mathscr{S}_2^0 = \mathscr{S}^0$ and if $\mathscr{S}^{\mu-1} \subseteq \mathscr{S}_j^{\mu-1}, j = 1, 2$, then

$$\begin{aligned}
\mathscr{S}^\mu &= \bigcap_{i=1}^{N}((\mathcal{S}_1 \cap \mathcal{S}_2) + (A_i^{-1}\mathscr{S}^{\mu-1} \cap \text{ Ker } C_i)) \\
&\subseteq \bigcap_{i=1}^{N}(\mathcal{S}_j + (A_i^{-1}\mathscr{S}^{\mu-1} \cap \text{ Ker } C_i)) \\
&\subseteq \bigcap_{i=1}^{N}(\mathcal{S}_j + (A_i^{-1}\mathscr{S}_j^{\mu-1} \cap \text{ Ker } C_i) = \mathscr{S}_j^\mu, \quad j = 1, 2
\end{aligned}$$

Therefore, $\mathscr{S}^n \subseteq \mathscr{S}_1^n \cap \mathscr{S}_2^n$, and consequently

$$\mathcal{S}_1 \cap \mathcal{S}_2 \subseteq \mathscr{S}^n \subseteq \mathscr{S}_1^n \cap \mathscr{S}_2^n = \mathcal{S}_1 \cap \mathcal{S}_2$$

and $\mathcal{S}_1 \cap \mathcal{S}_2 = \mathscr{S}^n$, so that the result follows from Theorem 5.4. ∎

Let $\mathcal{L} \subseteq \mathcal{X}$ be an arbitrary subspace and denote $\mathfrak{S}(\mathfrak{A}, \mathfrak{C}, \mathcal{L})$ (or simply $\mathfrak{S}(\mathcal{L})$) as the family of u.o.s. that contains $\mathcal{L}$, i.e.

$$\mathfrak{S}(\mathfrak{A}, \mathfrak{C}, \mathcal{L}) = \{\mathcal{S} : \mathcal{S} \in \mathfrak{S}(\mathfrak{A}, \mathfrak{C}, \mathcal{X}) \ \& \ \mathcal{L} \subseteq \mathcal{S}\}$$

As indicated previously, $\mathfrak{S}(\mathfrak{A}, \mathfrak{C}, \mathcal{X})$ refers to the class of all unobservability subspaces in $\mathcal{X}$.

Lemma 5.5 *For any given subspace $\mathcal{L} \subseteq \mathcal{X}$, $\mathfrak{S}(\mathfrak{A}, \mathfrak{C}, \mathcal{L})$ has an infimal element (denoted by $\mathcal{S}^* = \inf \mathfrak{S}(\mathfrak{A}, \mathfrak{C}, \mathcal{L})$).*

Proof: It follows that $\mathcal{X} \in \mathfrak{S}(\mathfrak{A}, \mathfrak{C}, \mathcal{L})$ and $\mathfrak{S}(\mathfrak{A}, \mathfrak{C}, \mathcal{L})$ is closed under intersection (Lemma 5.4). Therefore, it has an infimal element. ∎

The next algorithm provides a procedure for constructing $\mathcal{S}^*$.

Algorithm 5.3 *Let $\mathcal{W}^* = \inf \mathfrak{W}(\mathfrak{A}, \mathfrak{C}, \mathcal{L})$ and define the sequence $\mathcal{Z}^\mu$ according to*

$$\mathcal{Z}^0 = \mathcal{X}; \quad \mathcal{Z}^\mu = \bigcap_{i=1}^{N} (\mathcal{W}^* + (A_i^{-1} \mathcal{Z}^{\mu-1} \cap Ker\ C_i)) \tag{5.30}$$

Then $\mathcal{S}^* = \mathcal{Z}^k$, whenever $\mathcal{Z}^{k+1} = \mathcal{Z}^k$.

To analyze the above algorithm, note that the sequence $\mathcal{Z}^\mu$ is non-increasing and $\mathcal{Z}^{k+1} = \mathcal{Z}^k$ for $k \geq n - \dim(\mathcal{W}^*)$. Let $\mathcal{Z}^* = \mathcal{Z}^k$. According to Lemma 5.3, $\mathcal{Z}^* =<< \bigcap_{i=1}^N (\mathcal{W}^* + \text{Ker} C_j) | A_i + D_i C_i >>_{i \in \Psi}$ for some $(D_1, ..., D_N) \in \mathfrak{D}(\mathcal{W}^*)$. Using the same approach as in Lemma 5.1, one can obtain the maps H_j's such that

$$\mathcal{Z}^* =<< \bigcap_{i=1}^{N} (\mathcal{W}^* + \text{Ker} C_j) | A_i + D_i C_i >>_{i \in \Psi} \tag{5.31}$$

$$=<< \bigcap_{i=1}^{N} \text{Ker} H_j C_j | A_i + D_i C_i >>_{i \in \Psi} \tag{5.32}$$

and therefore, $\mathcal{Z}^*$ is an unobservability subspace according to Definition 5.6. Moreover, it follows that $\mathcal{L} \subseteq \mathcal{W}^* \subseteq \mathcal{Z}^*$ ($\mathcal{Z}^* \in \mathfrak{W}(\mathcal{L})$ and $\mathcal{W} = \inf \mathfrak{W}(\mathcal{L})$); hence $\mathcal{Z}^* \in \mathfrak{S}(\mathcal{L})$, and consequently $\mathcal{S}^* \subseteq \mathcal{Z}^*$.

On the other hand, according to Theorem 5.4, we have $\mathcal{S}^* = \mathcal{S}^n$ where

$$\mathcal{S}^0 = \mathcal{X}; \qquad \mathcal{S}^\mu = \bigcap_{i=1}^{N} (\mathcal{S}^* + (A_i^{-1} \mathcal{S}^{\mu-1} \cap \text{Ker}\ C_i)), \quad \mu \in \mathbf{n}$$

Since $\mathcal{W}^* \subseteq \mathcal{S}^*$, it can be shown by induction that $\mathcal{Z}^\mu \subseteq \mathcal{S}^\mu$, $\mu \in \mathbf{n}$. Indeed, $\mathcal{Z}^0 = \mathcal{S}^0$, and if $\mathcal{Z}^{\mu-1} \subseteq \mathcal{S}^{\mu-1}$, then

$$\begin{aligned}\mathcal{Z}^\mu &= \bigcap_{i=1}^{N} (\mathcal{W}^* + (A_i^{-1}\mathcal{Z}^{\mu-1} \cap \text{Ker } C_i)) \\ &\subseteq \bigcap_{i=1}^{N} (\mathcal{S}^* + (A_i^{-1}\mathcal{S}^{\mu-1} \cap \text{Ker } C_i)) = \mathcal{S}^\mu\end{aligned}$$

Consequently $\mathcal{Z}^* \subseteq \mathcal{S}^*$.

It follows from the above algorithm and Lemma 5.3 that $\mathfrak{D}(\mathcal{W}^*) \subseteq \mathfrak{D}(\mathcal{S}^*)$. Therefore, the maps D_i's for $\mathcal{S}^*$ can be found from $\mathcal{W}^*$ and once $\mathcal{S}^*$ is found from Algorithm 5.3, the maps H_i's can also be computed from $\mathcal{S}^*$ and equation (5.25).

The next lemma characterizes the relationship between $\mathcal{S}^*$ and $\mathcal{W}^*$ (similar to that of single mode systems as governed by equation (2.21)).

Lemma 5.6 *Given* $\mathcal{W}^* = \inf \mathfrak{W}(\mathfrak{A}, \mathfrak{C}, \mathcal{L})$, $\mathcal{S}^* = \inf \mathfrak{S}(\mathfrak{A}, \mathfrak{C}, \mathcal{L})$ *and* $(D_1, ..., D_N) \in \mathfrak{D}(\mathcal{W}^*)$, *we have*

$$\mathcal{S}^* =<< \bigcap_{i=1}^{N} (\mathcal{W}^* + Ker\ C_j)|A_i + D_iC_i >>_{i\in\Psi}$$

Proof: The proof is immediate from (5.31) and the fact that $\mathcal{Z}^* = \mathcal{S}^*$. ■

Finally, we introduce the notion of an outer MS-detectability of an unobservability subspace. As shown in the next section, outer MS-detectability is a necessary condition for designing a stable residual generator for the FDI problem.

Definition 5.7 *A subspace* $\mathcal{S} \in \mathfrak{S}(\mathfrak{A}, \mathfrak{C}, \mathcal{X})$ *is said to be outer MS-detectable if the factor system* $(\mathfrak{A} + \mathfrak{D}\mathfrak{C} : \mathcal{X}/\mathcal{S}, \mathfrak{H}\mathfrak{C} : \mathcal{X}/\mathcal{S}, \Pi)$ *is MS-detectable, where* $\mathfrak{H}\mathfrak{C} = \{H_1C_1, ..., H_NC_N\}$ *and* $\mathfrak{A} + \mathfrak{D}\mathfrak{C} = \{A_1 + D_1C_1, ..., A_N + D_NC_N\}$.

The next example illustrates how one can obtain $\mathcal{S}^*$ for a given Markovian jump system and it also summarizes the proposed methodology.

Example 5.1. Let $N = 2$ and

$$A_1 = \begin{bmatrix} 0 & 3 & 4 \\ 1 & 2 & 3 \\ 0 & 2 & 5 \end{bmatrix}, \quad A_2 = \begin{bmatrix} 1 & 2 & 4 \\ 2 & -1 & 2 \\ 0 & 1 & 4 \end{bmatrix}, C_1 = \begin{bmatrix} 0 & 1 & 0 \\ 0 & 0 & 1 \end{bmatrix}, C_2 = \begin{bmatrix} 1 & 1 & 0 \\ 0 & 0 & 1 \end{bmatrix},$$

$L = [-3, 1, 0]^\top$ and Π as in Example 1. First $\mathcal{W}^*$ can be found from the Algorithm 5.2 as

$$\mathcal{W}^* = \begin{bmatrix} -3 \\ 1 \\ 0 \end{bmatrix}$$

The unobservability subspace $\mathcal{S}^*$ is found from Algorithm 5.3 which is equal to $\mathcal{W}^*$. It should be noted that

$$\mathcal{S}_1^* = \begin{bmatrix} -3 & 1 \\ 1 & 0 \\ 0 & 0 \end{bmatrix}, \mathcal{S}_2^* = \begin{bmatrix} -3 \\ 1 \\ 0 \end{bmatrix}$$

where $\mathcal{S}_j^*$, $j = 1, 2$ is the infimal (C_j, A_j) unobservability subspace that contains $\mathcal{L} = \text{Im } L$ and $\mathcal{S}^* \subset \mathcal{S}_1^*$ and $\mathcal{S}^* \subset \mathcal{S}_2^*$.

5.4.2 A Geometric Approach to Fault Detection and Isolation of Discrete-Time MJS Systems

In this section the Fundamental Problem in Residual Generation (FPRG) is investigated for the discrete-time Markovian jump systems. This problem was originally considered for linear systems in [120] and was extended to nonlinear systems in [154]. The objective in this section is to generalize these results to Markovian jump systems (MJS).

Consider the following discrete-time Markovian jump system

$$\begin{aligned} x(k+1) &= A_{\lambda(k)}x(k) + B_{\lambda(k)}u(k) + L^1_{\lambda(k)}m_1(k) + L^2_{\lambda(k)}m_2(k) \\ y(k) &= C_{\lambda(k)}x(k), \quad x(0) = x_0, \quad \lambda(0) = i_0 \end{aligned} \tag{5.33}$$

where it is assumed that all the matrices are the same as in (5.11) and the Markov process $\lambda(k)$ is irreducible. The matrices $L^1_{\lambda(k)}$, $L^2_{\lambda(k)}$ represent the fault signatures and are monic and $m_i(k) \in \mathcal{M}_i \subset \mathcal{X}$, $i = 1, 2$ denote the fault modes. For sake of simplicity in the analysis and derivation of the results, we first consider the case with two faults. The more general case of multiple faults is considered at the end of this section. We denote the fault signatures that are associated with $\lambda(k) = i$ by L_i^1 and L_i^2. The fault modes together with the fault signatures can be used to model the effects of actuator faults, sensor faults and system faults on the dynamics of the system. For example, the effect of a fault in the i-th actuator may be represented by L_i^1 as the i-th column of B_i and if an actuator fails, then $m_1(k) = -u_i(k)$.

The FPRG problem is concerned with the design of a Markovian jump residual generator that is governed by the filter dynamics of the form

$$\begin{aligned} w(k+1) &= F_{\lambda(k)}w(k) - E_{\lambda(k)}y(k) + K_{\lambda(k)}u(k) \\ r(k) &= M_{\lambda(k)}w(k) - H_{\lambda(k)}y(k) \end{aligned} \tag{5.34}$$

where $w(k) \in \mathcal{F} \subset \mathcal{X}$ such that the response of $r(k)$ is affected by the fault mode $m_1(k)$ and is decoupled from $m_2(k)$ and if m_1 is identically zero then

$$\lim_{k\to\infty} \mathbb{E}||r(k)||^2 = 0$$

for any input signal $u(k)$. We can rewrite equations (5.33) and (5.34) as follows

$$\begin{bmatrix} x(k+1) \\ w(k+1) \end{bmatrix} = \begin{bmatrix} A_{\lambda(k)} & 0 \\ -E_{\lambda(k)}C_{\lambda(k)} & F_{\lambda(k)} \end{bmatrix} \begin{bmatrix} x(k) \\ w(k) \end{bmatrix} + \begin{bmatrix} B_{\lambda(k)} & L^2_{\lambda(k)} \\ K_{\lambda(k)} & 0 \end{bmatrix} \begin{bmatrix} u(k) \\ m_2(k) \end{bmatrix} + \begin{bmatrix} L^1_{\lambda(k)} \\ 0 \end{bmatrix} m_1(k)$$
$$r(k) = \begin{bmatrix} -H_{\lambda(k)}C_{\lambda(k)} & M_{\lambda(k)} \end{bmatrix} \begin{bmatrix} x(k) \\ w(k) \end{bmatrix} \tag{5.35}$$

Define the extended space $\mathcal{X}^e = \mathcal{X} \oplus \mathcal{F}$ and $\mathcal{U}^e = \mathcal{U} \oplus \mathcal{M}_2$, so that equation (5.35) can be expressed as

$$\begin{aligned} x^e(k+1) &= A^e_{\lambda(k)} x^e(k) + B^e_{\lambda(k)} u^e(k) + L^{e1}_{\lambda(k)} m_1(k) \\ r(k) &= H^e_{\lambda(k)} x^e(k) \end{aligned} \tag{5.36}$$

with $x^e(t) \in \mathcal{X}^e$ and $u^e \in \mathcal{U}^e$. In order to investigate the criteria for determining whether a nonzero $m_1(k)$ affects the residual signal $r(k)$, the notion of an input observability for the Markovian jump system is defined and formalized below.

Definition 5.8 *The input signal $m_1(t)$ is called input observable for the Markovian jump system (5.36) if L^{e1}_i, $i \in \Psi$ is monic and the image of L^{e1}_i's does not intersect with the unobservable subspace of system (5.36).*

Based on the above definition, the FPRG problem for system (5.33) can now be formally stated as the problem of designing the dynamical filter (5.34) such that

(a) r is decoupled from u^e, (5.37)

(b) m_1 is input observable in the augmented system (5.36), and (5.38)

(c) $\lim_{k\to 0} \mathbb{E}\{||r(k)||^2\} = 0$, for $m_1(k) = 0, \forall\, i_0 \in \Psi$ and $\forall x^e_0 \in \mathcal{X}^e$. (5.39)

We need to first derive a preliminary result for obtaining the solvability condition for the FPRG problem. The following embedding map $Q : \mathcal{X} \to \mathcal{X}^e$ is defined according to [120]

$$Qx = \begin{bmatrix} x \\ 0 \end{bmatrix} \tag{5.40}$$

where if $\mathcal{V} \subset \mathcal{X}^e$, we have

$$Q^{-1}\mathcal{V} = \{x | x \in \mathcal{X}, \quad \begin{bmatrix} x \\ 0 \end{bmatrix} \in \mathcal{V}\}$$

Our first result is the generalization of the Proposition 1 that was obtained in [120] to the Markovian jump systems.

Lemma 5.7 *Let $\mathcal{S}^e$ be the unobservable subspace of system* (5.36). *The unobservability subspace for system $(\mathfrak{A}, \mathfrak{C}, \Pi)$ is given by $Q^{-1}\mathcal{S}^e$.*

Proof: First, we show that $\mathcal{S} = Q^{-1}\mathcal{S}^e$ is conditioned invariant. Let $x \in \mathcal{S} \cap \text{Ker } C_i$, we need to show that $A_i x \in \mathcal{S}$. This follows by noting that

$$\begin{bmatrix} A_i x \\ 0 \end{bmatrix} = \begin{bmatrix} A_i & 0 \\ -E_i C_i & F_i \end{bmatrix} \begin{bmatrix} x \\ 0 \end{bmatrix} \in \mathcal{S}^e$$

since $\mathcal{S}^e$ is A_i^e-invariant, $i \in \Psi$. Therefore, $A_i x \in \mathcal{S}$ and $\mathcal{S}$ is conditioned invariant. Next if $x \in \mathcal{S}$, then $Qx \in \mathcal{S}^e$ and therefore

$$\begin{bmatrix} x \\ 0 \end{bmatrix} \in \bigcap_{i=1}^{N} \text{Ker } H_i^e$$

This shows that $H_i C_i x = 0$; and hence, $x \in \cap_{i=1}^N \text{Ker } H_i C_i$ and $\mathcal{S} \subseteq \cap_{i=1}^N \text{Ker } H_i C_i$. Finally, according to the definition of the unobservable subspace $\mathcal{S}^e$ (the largest A^e-invariant subspace in $\cap \text{Ker } H_i^e$), $\mathcal{S}$ is the largest conditioned invariant contained in $\cap_{i=1}^N \text{Ker } H_i C_i$, and therefore, $\mathcal{S} \in \mathfrak{S}(\mathfrak{A}, \mathfrak{C}, \Pi)$. ∎

We are now in a position to derive the solvability condition for the FPRG problem associated with the Markovian jump system (5.33).

Theorem 5.5 *The FPRG problem has a solution for the augmented MJS system* (5.36) *only if*

$$\mathcal{S}^* \bigcap \mathcal{L}_j^1 = 0, \quad j \in \Psi \tag{5.41}$$

where $\mathcal{S}^* = \inf \mathfrak{S}(\mathfrak{A}, \mathfrak{C}, \sum_{i=1}^N \mathcal{L}_i^2)$. On the other hand, if the above $\mathcal{S}^*$ exists such that it is also outer MS-detectable, then the FPRG problem is guaranteed to have a solution.

Proof: **(Only if part)** Let $\mathcal{S}^e$ be an unobservable subspace of system (5.36). To satisfy the condition (5.37), we should have $\mathcal{B}_i^e \subset \mathcal{S}^e, i \in \Psi$. Hence, $\mathcal{L}_{i2} \subset Q^{-1}\mathcal{B}_i^e \subset Q^{-1}\mathcal{S}^e = \mathcal{S}$ and by invoking Lemma 5.7, we obtain

$$\mathcal{S} \in \mathfrak{S}(\mathfrak{A}, \mathfrak{C}, \sum_{i=1}^{N} \mathcal{L}_{i2}) \tag{5.42}$$

For condition (5.38) to hold, according to the Definition 5.8, L_{i1}^e should be monic (which is already assumed to hold) and $\mathcal{L}_{i1}^e \cap \mathcal{S}^e = 0$, $i \in \Psi$. Thus

$$Q^{-1}(\mathcal{L}_{i1}^{e} \cap \mathcal{S}^{e}) = \mathcal{L}_{i1} \cap \mathcal{S} = 0, \quad i \in \Psi \tag{5.43}$$

Therefore, equations (5.42) and (5.43) hold only if equation (5.41) is true.
(if part): Given the unobservability subspace $\mathcal{S}^*$ which is outer MS-detectable, there exist output injection maps D_i's and measurement mixing maps H_i's such that $\mathcal{S}^* =<< \bigcap_{j=1}^{N} \text{Ker } H_j C_j | A_i + D_i C_i >>_{i \in \Psi}$, where H_j is the solution to the equation (5.25). Let P be the canonical projection of $\mathcal{X}$ on $\mathcal{X}/\mathcal{S}^*$ and $M_i, i \in \Psi$ be a unique solution to $M_i P = H_i C_i$ and $A_{0i} = (A_i + D_i C_i : \mathcal{X}/\mathcal{S}^*)$, where $P(A_i + D_i C_i) = A_{0i} P$, $i \in \Psi$. Due to the fact that $\mathcal{S}^*$ is assumed to be outer MS-detectable, there exist G_i, $i \in \Psi$ such that $(\mathfrak{A}_{0_G}, \Pi)$ is MS-stable, where $\mathfrak{A}_{0_G} = \{A_{01} + G_1 M_1, ..., A_{0N} + G_N M_N\}$. Let us define $F_i = A_{0i} + G_i M_i$, $E_i = P(D_i + P^{-r} G_i H_i)$, $K_i = PB_i$, $i \in \Psi$, and $e(k) = w(k) - Px(k)$. By using equation (5.34) one obtains

$$\begin{aligned}
e(k+1) =& F_i w(k) - E_i y(k) + K_i u(k) \\
& - P(A_i x(k) + B_i u(k) + L_{i1} m_1(k) + L_{i2} m_2(k)) \\
=& F_i w(k) - P L_{i1} m_1(k) - \\
& P(A_i + D_i C_i) x(k) - G_i H_i C_i x(k) \\
=& F_i w(k) - P L_{i1} m_1(k) - A_{0i} P x(k) - G_i M_i P x(k) \\
=& F_i e(k) - P L_{i1} m_1(k)
\end{aligned}$$

Note that $PL_{i2} = 0, i \in \Psi$, since $\mathcal{L}_{i2} \in \mathcal{S}^*, i \in \Psi$. Also

$$r(k) = M_i w(k) - H_i y(k) = M_i w(k) - H_i C_i x(k) = M_i e(k)$$

Consequently, the error dynamics can be re-written according to

$$\begin{aligned}
e(k+1) &= F_{\lambda(k)} e(k) - P L_{\lambda(k)}^{1} m_1(k) \\
r(k) &= M(\lambda) e(k)
\end{aligned} \tag{5.44}$$

It follows that the fault mode $m_2(k)$ does not affect the residual signal $r(k)$ and since the dynamics (5.44) is observable, condition (5.38) also holds. Moreover, for $m_1(k) = 0$ the system is MS-stable and condition (5.39) holds. ■

To conclude this section, we now consider a discrete-time Markovian jump system that has multiple faults and that is governed by the following dynamical system

$$\begin{aligned}
x(k+1) &= A_{\lambda(k)} x(k) + B_{\lambda(k)} u(k) + \sum_{j=1}^{P} L_{\lambda(k)}^{j} m_j(k) \\
y(k) &= C_{\lambda(k)} x(k) \qquad x(0) = x_0, \quad \lambda(0) = i_0
\end{aligned} \tag{5.45}$$

where all the matrices are the same as in the dynamical model (5.33), $L_{\lambda(k)}^{j}$, $j \in \mathbf{k}$ denote the fault signatures, and $m_j(k) \in \mathcal{M}_j$, $j \in \mathbf{P}$ denote the fault modes. We denote the fault signatures associated with $\lambda(k) = i$ by $L_i^j, i \in \Psi, j \in \mathbf{P}$.

The Structured Fault Detection and Isolation Problem (SFDIP) (Section 2.1) for the Markovian jump system (5.45) is now defined as the problem of generating p residual signals $r_l(k), l \in \mathbf{p}$ based on a given coding set $\Omega_i, i \in \mathbf{P}$, from the following Markovian jump detection filters

$$\begin{aligned} w_l(k+1) &= F^l_{\lambda(k)} w_l(k) - E^l_{\lambda(k)} y(k) + K^l_{\lambda(k)} u(k) \\ r_l(k) &= M^l_{\lambda(k)} w_l(k) - H^l_{\lambda(k)} y(k), \qquad l \in \mathbf{p} \end{aligned} \tag{5.46}$$

such that the residuals $r_l(k)$ for $l \in \Omega_i$ are sensitive to a fault of the i-th component, and the other residuals $r_\alpha(k)$ for $\alpha \in \mathbf{p} - \Omega_i$ are insensitive to this fault. The solvability condition for the SFDIP problem is obtained by invoking the solvability condition that was developed earlier for the FPRG problem as follows:

Theorem 5.6 *The SFDIP problem has a solution for system* (5.45) *only if*

$$\mathcal{S}^*_{\Gamma_j} \bigcap \mathcal{L}^l_i = 0, \quad i \in \Psi, l \in \Gamma_j \tag{5.47}$$

where $\mathcal{S}^*_{\Gamma_j} = \inf \mathfrak{S}(\mathfrak{A}, \mathfrak{C}, \sum_{v=1}^N \sum_{l \notin \Gamma_j} \mathcal{L}^l_v)$, $j \in \mathbf{p}$. *On the other hand, if the above* $\mathcal{S}^*_{\Gamma_j}, j \in \mathbf{p}$ *exist such that they are outer MS-detectable, the EFPRG problem is then guaranteed to have a solution.*

Proof: The proof is immediate by following along the same steps that are invoked as in the proof of Theorems 5.5 and 2.3. ■

5.4.3 An H_∞-based Fault Detection and Isolation Design Strategy

In this section, we consider a discrete-time Markovian jump system that is subjected to both external input and output measurement disturbances and noise and that is governed by

$$\begin{aligned} x(k+1) &= A_{\lambda(k)} x(k) + B_{\lambda(k)} u(k) + \sum_{j=1}^{P} L^j_{\lambda(k)} m_j(k) + B^d_{\lambda(k)} d(k) \\ y(k) &= C_{\lambda(k)} x(k) + D^d_{\lambda(k)} d(k) \end{aligned} \tag{5.48}$$

where all the matrices are defined as in (5.45). The signal $d(k) \in \mathbb{R}^p$ represents an unknown disturbance input and output measurement noise. We denote the disturbance matrices $B^d_{\lambda(k)}$ and $D^d_{\lambda(k)}$ associated with $\lambda(k) = i$ by B^d_i and D^d_i, $i \in \Psi$. It is further assumed that the disturbance input $d(t)$ belongs to $\mathscr{L}_2$, i.e.,

$$||d||_2 = \Big(\sum_{k=0}^{\infty} d^\top(k)d(k)\Big)^{1/2} < \infty$$

Based on the above representation of the Markovian jump system (5.48), an H_∞-based Structured Fault Detection and Isolation Problem (H_∞-SFDIP) is formulated which is concerned with the design of a set of detection filters (5.46) that generate p residuals $r_j(k)$ based on given coding sets $\Omega_i, i \in \mathbf{P}$ such that the residuals $r_j(t)$ for $j \in \Omega_i$ are sensitive to a fault of the i-th component, and the other residuals $r_\alpha(t)$ for $\alpha \in \mathbf{p} - \Omega_i$ are insensitive to this fault and

$$||r_l||_{2,E}^2 = \mathbb{E}\Big\{\sum_{k=0}^{\infty} r_l^\top(k)r_l(k)|(x_0, i_0)\Big\} < \gamma^2||d||_2, \quad l \in \mathbf{p} \tag{5.49}$$

for all $d(k) \in \mathscr{L}_2$, where $\gamma > 0$ is the prescribed level of disturbance attenuation.

Below, we first present preliminary results on disturbance attenuation of Markovian jump systems.

Lemma 5.8 ([162]) *Consider the system* (5.48) *with* $u(k) = 0$ *and* $m_j(k) = 0, j \in \boldsymbol{P}$. *Let* $\gamma > 0$ *be a given scalar, then the system is mean square stable when* $d = 0$ *and under zero initial conditions satisfies*

$$||y||_{2,E} < \gamma||d||_2 \tag{5.50}$$

if there exist matrices $R_i > 0, i \in \Psi$ *such that the following LMIs:*

$$\begin{bmatrix} A_i^\top \bar{R}_i A_i - R_i & A_i^\top \bar{R}_i B_i^d & C_i^\top \\ * & -\gamma^2 I + B_i^{d^\top} \bar{R}_i B_i^d & D_i^{d^\top} \\ * & * & -I \end{bmatrix} < 0 \tag{5.51}$$

hold for $i \in \Psi$ *where* $\bar{R}_i = \sum_{j=1}^{N} \pi_{ij} R_j$. *Moreover, the LMI* (5.51) *is equivalent to*

$$\begin{bmatrix} -R_i & A_i^\top \bar{R}_i & 0 & C_i^\top \\ * & -\bar{R}_i & \bar{R}_i B_i^d & 0 \\ * & * & -\gamma^2 I & D_i^{d^\top} \\ * & * & * & -I \end{bmatrix} < 0 \tag{5.52}$$

for $i \in \Psi$.

Lemma 5.9 ([51]) *The LMI* (5.52) *with* $R_i > 0$, $i \in \Psi$ *is feasible if and only if there exist matrices* $R_i > 0$ *and* $\mathscr{G}_i$, $i \in \Psi$ *such that*

$$\begin{bmatrix} -R_i & A_i^\top \mathscr{G}_i^\top & 0 & C_i^\top \\ * & -\mathscr{G}_i^\top - \mathscr{G}_i + \bar{R}_i & \mathscr{G}_i B_i^d & 0 \\ * & * & -\gamma^2 I & D_i^{d^\top} \\ * & * & * & -I \end{bmatrix} < 0$$

We are now in a position to derive the sufficient conditions for determining the solvability of the H_∞-SFDIP problem for discrete-time MJS systems.

Theorem 5.7 *The H_∞-SFDIP problem defined by expressions* (5.49) *and* (5.46) *has a solution for the Markovian jump system* (5.48) *if there exist unobservability subspaces*

$$\mathcal{S}_{\Gamma_l}^* = \inf \mathfrak{S}(\mathfrak{A}, \mathfrak{C}, \sum_{v=1}^{N} \sum_{j \notin \Gamma_l} \mathcal{L}_v^j), \quad l \in \mathbf{p} \tag{5.53}$$

such that $\mathcal{S}_{\Gamma_l}^* \bigcap \mathcal{L}_i^j = 0$, $i \in \Psi, j \in \Gamma_l$ as well as the matrices T_i^l, $\mathscr{G}_i^l$ and positive-definite matrices R_i^l, $i \in \Psi$, $l \in \mathbf{p}$ such that

$$\begin{bmatrix} -R_i^l & A_i^{l^\top} \mathscr{G}_i^l + M_i^{l^\top} T_i^{l^\top} & 0 & M_i^{l^\top} \\ * & -\mathscr{G}_i^{l^\top} - \mathscr{G}_i^l + \bar{R}_i^l & \mathscr{G}_i^l B_{li}^d - T_i^l H_i^l D_i^d & 0 \\ * & * & -\gamma^2 I & D_{li}^{d^\top} \\ * & * & * & -I \end{bmatrix} < 0 \tag{5.54}$$

where $\bar{R}_i^l = \sum_{j=1}^N \pi_{ij} R_j^l$ and P_l is the canonical projection of $\mathcal{X}$ on $\mathcal{X}/\mathcal{S}_{\Gamma_l}^*$, $B_{li}^d = -P_l B_i^d - P_l D_i^l D_i^d$, $D_{li}^d = -H_i^l D_i^d$, the pairs (M_i^l, A_i^l), $i \in \Psi$, $l \in \mathbf{p}$ are the factor system of the pairs $(C_i, A_i), i \in \Psi$ on $\mathcal{X}/\mathcal{S}_{\Gamma_l}^*$ and H_i^l is the solution to Ker $H_i^l C_i = \mathcal{S}_{\Gamma_l}^* +$ Ker C_i.

Proof: Given the common unobservability subspaces $\mathcal{S}_{\Gamma_l}^*$, there exist output injection maps D_i^l and measurement mixing maps H_i^l $i \in \Psi$, $l \in \mathbf{p}$ such that

$$\mathcal{S}_{\Gamma_l}^* =<< \text{Ker } H_i^l C_i | A_i + D_i^l C_i >>_{i \in \Psi}$$

where H_i^l is the solution to Ker $H_i^l C_i = \mathcal{S}_l^* +$ Ker C_i. Let M_i^l be a unique solution to $M_i^l P_l = H_i^l C_i$ and

$$A_i^l = (A_i + D_i^l C_i : \mathcal{X}/\mathcal{S}_{\Gamma_l}^*)$$

where $P_l(A_i + D_i^l C_i) = A_i^l P_l$. Let T_i^l and $\mathscr{G}_i^l$ represent the solution to the inequality (5.54) and define $G_i^l = \mathscr{G}_i^{l^{-1}} T_i^l$ and $F_i^l = A_i^l + G_i^l M_i^l$, $E_i^l = P_l(D_i^l + P_l^{-r} G_i^l H_i^l)$. Let $K_i^l = P_l B_i$. Define $e_l(k) = w_l(k) - P_l x(k)$, then by using (5.34) one gets

$$\begin{aligned}
e_l(k+1) =& F_i^l w_l(k) - E_i^l y(k) + K_i^l u(k) \\
& - P_l(A_i x(k) + B_i u(k) + B_i^d d(k) + \sum_{j=1}^{L} L_i^j m_j(k)) \\
=& F_i^l w_l(k) - \sum_{j\in\Gamma_l} P_l L_i^j m_j(k) - P_l B_i^d d(k) \\
& - P_l(A_i + D_i^l C_i)x(k) - G_i^l H_i^l C_i x(k) - E_i^l D_i^d d(k) \\
=& F_i^l w_l(k) - \sum_{j\in\Gamma_l} P_l L_i^j m_j(k) + (B_{li}^d \\
& - G_i^l H_i^l D_i^d) d(k) - A_i^l P_l x(k) - G_i^l M_i^l P_l x(k) \\
=& (A_i^l + G_i^l M_i^l) e_l(k) - \sum_{j\in\Gamma_l} P_l L_i^j m_j(k) + (B_{li}^d - G_i^l H_i^l D_i^d) d(k)
\end{aligned}$$

Note that $P_l L_i^j = 0, i \in \Psi, j \notin \Gamma_l$, since $\mathcal{L}_i^j \in \mathcal{S}_{\Gamma_l}^*, j \notin \Gamma_l$. Also

$$\begin{aligned}
r_l(t) &= M_i^l w_l(k) - H_i^l y(k) = M_i^l w_l(k) - H_i^l C_i x(k) - H_i^l D_i^d d(k) \\
&= M_i^l e_l(k) + D_{li}^d d(k)
\end{aligned}$$

Consequently, the error dynamics can be written as

$$\begin{aligned}
e_l(k+1) =& (A_i^l + G_i^l M_i^l) e_l(k) - \sum_{j\in\Gamma_l} P_l L_i^j m_j(k) + (B_{li}^d - G_i^l H_i^l D_i^d) d(k) \\
r_l(k) =& M_i^l e_l(k) + D_{li}^d d(k)
\end{aligned} \tag{5.55}$$

By using the Lemma 5.8 and the inequality (5.52), it follows that the inequality (5.49) holds. Moreover, from the error dynamics (5.55), it follows that $r_l(k)$ is only affected by $L_{\lambda(k)}^j, j \in \Gamma_l$ and is decoupled from the other fault signatures. ■

After constructing the residual signals $r_l(k), l \in \mathbf{p}$, the last step for a successful fault detection and isolation is the residual evaluation stage which includes determining the evaluation functions J_{r_l} and the thresholds J_{th_l}. In this section, evaluation functions and thresholds are selected as

$$J_{r_l}(k) = \sum_{k-k_0}^{k} r_l^\top(k) r_l(k), \quad l \in \mathbf{p} \tag{5.56}$$

$$J_{th_l} = \sup_{d\in\mathfrak{L}_2, m_j=0, j\in\Gamma_l} E(J_{r_l}), \quad l \in \mathbf{p} \tag{5.57}$$

where k_0 is the length of the evaluation window. According to the above thresholds and evaluation functions, the occurrence of a fault can be detected and isolated by using the following decision logics

$$J_{r_l} > J_{th_l} \quad \forall l \in \Omega_j \Longrightarrow m_j \neq 0, \quad j \in \mathbf{P} \tag{5.58}$$

5.4.4 The FDI Scheme for Formation Flight of Satellites

In order to investigate the effects of packet dropout on the FDI scheme for formation flight of four satellites (Section 3.3.3), two different topologies (neighboring sets) are considered. The first one is the same as the network topology that is considered in Section 3.3.3, namely $N_1 = \{2\}$, $N_2 = \{1,3\}$, $N_3 = \{4\}$ and $N_4 = \{1\}$. It can be verified that the SFDIP problem has a solution for the same coding sets as in Section 3.3.3, and hence the packet dropout does not change the detectability index of the actuator fault signatures of the satellites. This is due to the fact that the graph of the network remains weakly connected when all the arcs that leave node i are deleted for each $i \in \mathbf{N}$.

The second topology that is considered is specified as $N_1 = \{2\}$, $N_2 = \{3\}$, $N_3 = \{4\}$ and $N_4 = \{2\}$. It can be verified that the SFDIP problem has a solution with the same coding sets as in the first topology when the communication links are considered to be ideal (without time-delay and packet dropout, i.e. system (5.2)). However, in the presence of imperfect communication links (system (5.7)), the SFDIP problem does not have a solution for this coding sets, and hence the detectability index of the actuator fault signatures is less than the ideal case. This is due to the fact that the graph of the network does not remain weakly connected when the arc leaving node 1 to node 2 is deleted. In other words, when the measurement from the satellite 1 is lost, there exists no way to recover this information.

Next, we try to find a suitable coding sets for the above network. Due to the lack of measurement from satellite 1 when there is a packet dropout in the communication link between this satellite and the central FDI unit, actuator faults in satellite 1 cannot be decoupled. In other words, we have $\mathcal{L}_{11} \subset \mathcal{S}^*(\mathcal{L}_{12})$ and $\mathcal{L}_{12} \subset \mathcal{S}^*(\mathcal{L}_{11})$. Hence, these two fault signatures are considered together as a subfamily $FL_1 = \{\mathcal{L}_{11}, \mathcal{L}_{12}\}$. Now, we consider the generalized residual set for the 7 fault signatures FL_1 and $\mathcal{L}_{kj}$, $k \in \{2,3,4\}$, $j \in \{1,2\}$. In other words, we consider the following coding sets $\Omega_{FL_1} = \{2,3,4,5,6,7\}$, $\Omega_{21} = \{1,3,4,5,6,7\}$, $\Omega_{22} = \{1,2,4,5,6,7\}$, $\Omega_{31} = \{1,2,3,5,6,7\}$, $\Omega_{32} = \{1,2,3,4,6,7\}$, $\Omega_{41} = \{1,2,3,4,5,7\}$ and $\Omega_{41} = \{1,2,3,4,5,6\}$. It can be verified that the SFDIP problem has a solution for these coding sets, and hence one can detect and isolate single fault in the actuators of satellites 2,3 and 4 but actuator faults of satellite 1 cannot be isolated and only occurrence of a fault in this satellite can be detected and isolated.

For simulations a 10% packet dropout rate is considered for all the communication links, and hence the following transition probability is considered for the Markovian jump model of the satellite formation

$$\Pi = \begin{bmatrix} 0.5 & 0.1 & 0.1 & 0.1 & 0.1 \\ 0.95 & 0.05 & 0 & 0 & 0 \\ 0.95 & 0 & 0.05 & 0 & 0 \\ 0.95 & 0 & 0 & 0.05 & 0 \\ 0.95 & 0 & 0 & 0 & 0.05 \end{bmatrix}$$

The length of the evaluation window is selected as $k_0 = 50$ (5 seconds). By considering the worst case analysis of the residuals corresponding to the healthy operation of the satellites that are subject to measurement noise, threshold values $J_{th_1} = J_{th_2} = 0.002$, $J_{th_3} = 0.001$, $J_{th_4} = J_{th_5} = J_{th_6} = J_{th_7} = 0.0006$ are selected for the residual signals for fault detection and isolation logic evaluation and analysis.

Two fault scenarios are considered for simulation studies. In the first one, a lock-in-place fault is injected at the first actuator of satellite 1 at $t = 20$ seconds where $u_{11} = -1$. Figure 5.3 shows the residual evaluation functions that are generated by using the detection filters associated with the considered fault scenario. As shown in this figure, the injected fault can be detected and isolated among the satellites based on the coding set Ω_{FL_1}. However, one cannot isolate between the actuators of satellite 1. In the second fault scenario, a lock-in-place fault is injected at the second actuator of satellite 2 at $t = 25$ seconds where $u_{22} = 0.5$. Figure 5.4 depicts the residual evaluation functions corresponding to this fault scenario. Based on this figure and the coding sets Ω_{22}, this fault can be detected and isolated.

5.5 Continuous-time Markovian Jump Systems (MJS)

Consider the following continuous-time Markovian jump system (MJS)

$$\begin{aligned} \dot{x}(t) &= A(\lambda(t))x(t) + B(\lambda(t))u(t) \\ y(t) &= C(\lambda(t))x(t), \quad x(0) = x_0, \quad \lambda(0) = i_0 \end{aligned} \tag{5.59}$$

where $x \in \mathcal{X}$ is the state of the system with dimension n; $u \in \mathcal{U}, y \in \mathcal{Y}$ are input and output signals with dimensions m and q, respectively; and $\{\lambda(t), t \geq 0\}$ is a continuous-time irreducible Markov process taking values in the finite set $\Psi = \{1, ..., N\}$. For a reducible Markov process, refer to the work [134]. The Markov process describes the switching between the different system modes and its evolution is governed by the following probability transitions

$$\mathbb{P}\{\lambda(t+h) = j | \lambda(t) = i\} = \begin{cases} \pi_{ij}h + o(h) & \text{when } \lambda(t) \text{ jumps from } i \text{ to } j \\ 1 + \pi_{ii}h + o(h), & \text{otherwise} \end{cases}$$

where π_{ij} is the transition rate from mode i to mode j, with $\pi_{ij} \geq 0$ when $i \neq j$, $\pi_{ii} = -\sum_{j=1, j\neq i}^{N} \pi_{ij}$, and $o(h)$ is a function that satisfies $lim_{h\to 0} \frac{o(h)}{h} =$

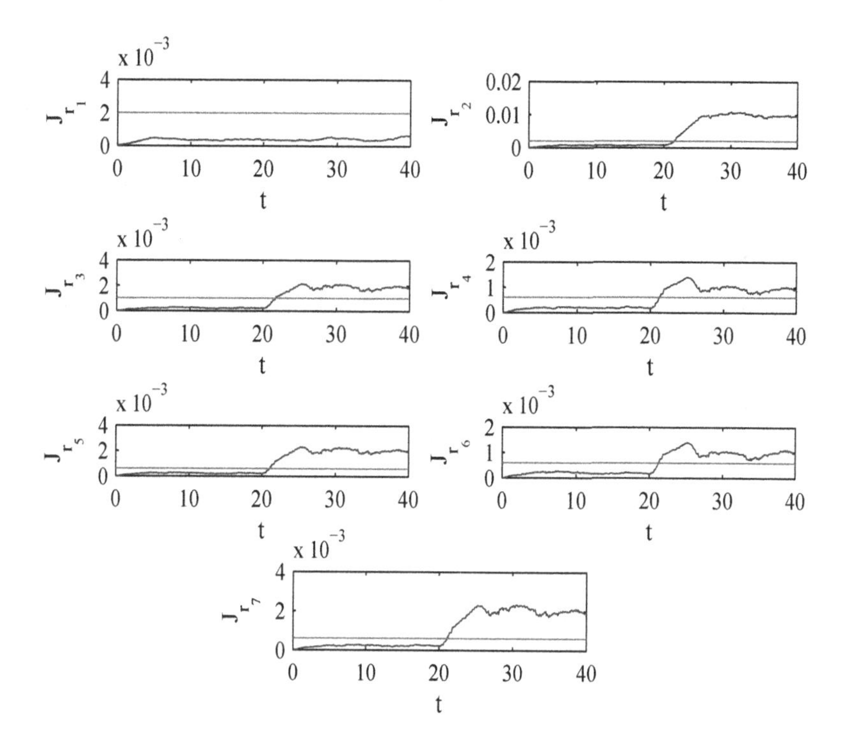

Fig. 5.3 The residual evaluation functions corresponding to a fault in the first actuator of satellite 1.

0. The matrices $A(\lambda(t))$, $B(\lambda(t))$ and $C(\lambda(t))$ are known constant matrices for all $\lambda(t) = i \in \Psi$. For simplicity, we denote the matrices associated with $\lambda(t) = i$ by $A(\lambda(t)) = A_i$, $B(\lambda(t)) = B_i$, and $C(\lambda(t)) = C_i$. Furthermore, the MJS system (5.59) is represented by $(\mathfrak{A}, \mathfrak{B}, \mathfrak{C}, \Pi)$, where $\mathfrak{A} = (A_1, ..., A_N)$, $\mathfrak{B} = (B_1, ..., B_N)$, $\mathfrak{C} = (C_1, ..., C_N)$ and $\Pi = [\pi_{ij}],\ i, j \in \Psi$.

5.5.1 Unobservable and Unobservability Subspaces

Similar to the presentation in Section 5.4.1, in this section a geometric definition for the unobservable subspace of continuous-time Markovian jump systems (MJS) is introduced. The notion of unobservability subspace is then formalized for continuous-time MJS systems that are governed by (5.59). We first start with the definition of weak observability for the Markovian jump system (5.59) with $B_i = 0, i \in \Psi$.

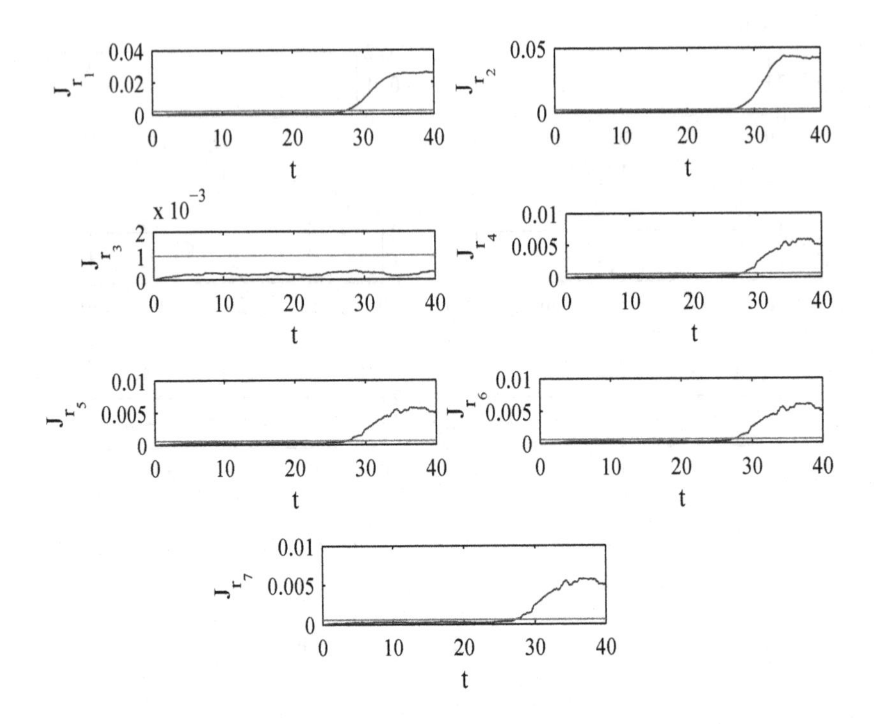

Fig. 5.4 The residual evaluation functions corresponding to a fault in the second actuator of satellite 2.

Definition 5.9 ([45]) *The system* $(\mathfrak{A}, \mathfrak{C}, \Pi)$ *is said to be weakly (W-) observable when there exist* $t_d \geq 0$ *and* $\gamma > 0$ *such that* $W^{t_d}(x, i) \geq \gamma |x|^2$ *for each* $x \in \mathcal{X}$ *and* $i \in \Psi$ *where*

$$W^{\top}(x,i) = \mathbb{E}\Big\{ \int_0^{\top} x^{\top}(\tau) C_{\lambda(\tau)}^{\top} C_{\lambda(\tau)} x(\tau) d\tau | x(0) = x, \lambda_0 = i \Big\} \tag{5.60}$$

In [45], a collection of matrices $\mathcal{O} = \{\mathcal{O}_1, ..., \mathcal{O}_N\}$ is introduced for testing the W-observability of Markovian jump systems according to the following procedure. Let $O_i(0) = C_i^{\top} C_i,\ i \in \Psi$ and define the sequence of matrices as

$$O_i(k) = A_i^{\top} O_i(k-1) + O_i(k-1) A_i + \sum_{j=1}^{N} \pi_{ij} O_j(k-1), \quad k > 0, i \in \Psi \tag{5.61}$$

The matrix $\mathcal{O}_i$ is defined according to

$$\mathcal{O}_i = [O_i(0)\ O_i(1) \cdots O_i(n^2N-1)]^\top \tag{5.62}$$

Theorem 5.8 ([45]) *The MJS system (5.59) is W-observable if and only if $\mathcal{O}_i$ has a full rank for each $i \in \Psi$.*

It is also known that system (5.59) is W-observable if each pair $(A_i, C_i), i \in \Psi$ is observable. However, this condition is not necessary [45]. By considering the above definition of W-observability, one can define the set of unobservable states as follows.

Definition 5.10 *A state x is said to be unobservable if $W^\top(x,i) = 0$ for all $t \geq 0$ and $i \in \Psi$.*

Let $\mathcal{Q}$ denotes the unobservable set of the MJS system (5.59), i.e.,

$$\mathcal{Q} = \{x | W^\top(x,i) = 0, \forall i \in \Psi, t \geq 0\} \tag{5.63}$$

It is shown in [45] that for an irreducible Markov processes, $\mathcal{N}\{\mathcal{O}_i\} = \mathcal{N}\{\mathcal{O}_j\}$, $i,j \in \Psi$, $i \neq j$ and $\mathcal{Q} = \mathcal{N}\{\mathcal{O}_i\}$. Therefore, $\mathcal{Q}$ is a subspace of $\mathcal{X}$ and is called the unobservable subspace of the Markovian jump system (5.59). The theorem introduced below characterizes a geometric property of $\mathcal{Q}$.

Theorem 5.9 *An unobservable subspace Q for the system $(\mathfrak{A}, \mathfrak{C}, \Pi)$ with an irreducible Markov process is the largest A_i-invariant ($i \in \Psi$) that is contained in $\mathcal{K} = \bigcap_{i=1}^N$ Ker C_i.*

Proof: It follows from the above discussion that $\mathcal{Q} \subseteq$ Ker C_i, $i \in \Psi$, and hence $\mathcal{Q} \subseteq \mathcal{K}$. Let $x \in \mathcal{Q}$. The goal is to show that $A_i x \in \mathcal{Q}$ for all $i \in \Psi$ (i.e. $\mathcal{Q}$ is A_i-invariant, $i \in \Psi$). Since $x \in \mathcal{N}\{O_i(k-1)\}$ and $x \in \mathcal{N}\{O_i(k)\}$, $i \in \Psi$, then

$$\begin{aligned} O_i(k)x &= A_i^\top O_i(k-1)x + O_i(k-1)A_i x + \sum_{j=1}^N \pi_{ij} O_j(k-1)x \\ &= O_i(k-1)A_i x = 0 \end{aligned}$$

Hence, $A_i x \in \mathcal{N}\{\mathcal{O}_i(k-1)\}$ and $\mathcal{Q}$ is A_i-invariant for all $i \in \Psi$.

Next, we show that Q is the largest A_i-invariant ($i \in \Psi$) that is contained in $\mathcal{K}$. Let $\mathcal{V}$ be an A_i-invariant ($i \in \Psi$) subspace that is contained in $\mathcal{K}$. Clearly, $\mathcal{V} \subseteq \mathcal{N}\{O_i(0)\}$, $i \in \Psi$. Let $\mathcal{V} \subseteq \mathcal{N}\{O_i(k-1)\}$, $i \in \Psi$ and $x \in \mathcal{V}$, then

$$O_i(k)x = A_i^\top O_i(k-1)x + O_i(k-1)A_i x + \sum_{j=1}^N \pi_{ij} O_j(k-1)x = 0$$

since $O_i(k-1)x = 0, i \in \Psi$ and $A_i x \in \mathcal{V}$ ($\mathcal{V}$ is A_i-invariant). Hence, $\mathcal{V} \subseteq \mathcal{N}\{O_i(k)\}$, $i \in \Psi$ and $\mathcal{V} \subseteq \mathcal{Q}$. This shows that $\mathcal{Q}$ contains all the subspaces that are A_i-invariant ($i \in \Psi$) and is contained in $\mathcal{K}$. ■

The same algorithm as in Algorithm 5.1 can be utilized for obtaining an unobservable subspace of the continuous-time Markovian jump system (5.59). Moreover, similar definition of conditioned invariant and unobservability subspaces for discrete-time MJS can be applied to the continuous-time MJS (Definitions 5.5 and 5.6, respectively), and the same algorithm can be used for obtaining such subspaces (Algorithms 5.2 and 5.3, respectively).

Next, we present definitions for stability and detectability of MJS systems.

Definition 5.11 ([45]) *The system $(\mathfrak{A}, \Pi)$ is mean square (MS) stable if for each $x_0 \in \mathcal{X}$ and $i_0 \in \Psi$,*

$$\lim_{t \to \infty} \mathbb{E}\{||x(t)||^2\} = 0 \tag{5.64}$$

MS-detectability of continuous-time MJS is defined similarly to the discrete-time MJS as follows.

Definition 5.12 ([45]) *We say that $(\mathfrak{A}, \mathfrak{C}, \Pi)$ is MS-detectable when there exists $G = \{G_1, ..., G_N\}$ of appropriate dimension for which $(\mathfrak{A}_G, \Pi)$ is MS-stable, where $\mathfrak{A}_G = \{A_1 + G_1C_1, ..., A_N + G_NC_N\}$.*

The following computational linear matrix inequalities can be used for testing the MS-detectability of a continuous-time MJS system [45]. In other words, the MS-detectability of $(\mathfrak{A}, \mathfrak{C}, \Pi)$ is equivalent to the feasibility of the set

$$A_i^\top X_i + X_iA_i + C_i^\top L_i^\top + L_iC_i + \sum_{j=1}^{N} \pi_{ij}X(j) < 0, \quad i = 1, ..., N \tag{5.65}$$

in the unknowns $X_i > 0$ and L_i with appropriate dimensions. In [45] an example is given that is W-observable but not MS-detectable. The example below illustrates the results that we have developed so far for the MJS systems. It also shows that an MJS system could be MS-detectable or not depending on the mode transition matrix.

Example 5.2. Let $N = 2$ and set

$$A_1 = \begin{bmatrix} 1\,0 \\ 0\,1 \end{bmatrix}, A_2 = \begin{bmatrix} 2\,3 \\ 0\,5 \end{bmatrix}, C_1 = \begin{bmatrix} 0\,0 \end{bmatrix}, C_2 = \begin{bmatrix} 0\,1 \end{bmatrix}; \Pi = \begin{bmatrix} -1 & 1 \\ 1 & -1 \end{bmatrix}$$

From the expression (5.62), one can evaluate rank($\mathcal{O}_1$)=rank($\mathcal{O}_2$)=1 and Theorem 5.8 ensures that this system is not W-observable. Using the Algorithm 5.1, the unobservable subspace $\mathcal{Q}$ is found to be

$$\mathcal{Q} = \begin{bmatrix} 1 \\ 0 \end{bmatrix}$$

The factor system for the above system can be found as $(A_1 : \mathcal{X}/\mathcal{S}) = 1$, $(A_2 : \mathcal{X}/\mathcal{S}) = 5$, $(C_1 : \mathcal{X}/\mathcal{S}) = 0$, and $(C_2 : \mathcal{X}/\mathcal{S}) = 1$, which is clearly

W-observable, but according to (5.65) it is not MS-detectable. However, corresponding to the probability transition matrix

$$\Pi = \begin{bmatrix} -10 & 10 \\ 3 & -3 \end{bmatrix}$$

the above factor system is both W-observable and MS-detectable.

The above example shows that W-observability is a structural property of the MJS system with irreducible Markov processes which depends on only the matrices A_i's and C_i's (this also follows from Theorem 5.9), but the MS-detectability depends furthermore on the mode transition matrix Π.

Remark 5.3. It should be emphasized that the proposed Theorem 5.9 and Algorithm 5.1 do indeed provide a less computationally intensive method for verifying the W-observability of the Markovian jump systems in comparison with the results in [45], where one needs to obtain the N matrices $\mathcal{O}_i \in \mathcal{R}^{n(n^2N)\times n}$.

We are now in a position to formally introduce the fundamental problem in residual generation (FPRG) for the continuous-time Markovian jump systems.

5.5.2 A Geometric Approach to the Fault Detection and Isolation of Continuous-Time MJS Systems

Similar to the presentation in Section 5.4.2, in this section the Fundamental Problem in Residual Generation is now investigated for the continuous-time Markovian jump systems. Consider the following Markovian jump system (MJS)

$$\begin{aligned} \dot{x}(t) &= A(\lambda(t))x(t) + B(\lambda(t))u(t) + L_1(\lambda(t))m_1(t) + L_2(\lambda(t))m_2(t) \\ y(t) &= C(\lambda(t))x(t), \quad x(0) = x_0, \quad \lambda(0) = i_0 \end{aligned} \tag{5.66}$$

where it is assumed that all the matrices are the same as in (5.59) and the Markov process $\lambda(t)$ is irreducible. The matrices $L_1(\lambda(t))$ and $L_2(\lambda(t))$ represent the fault signatures and are monic and $m_i(t) \in \mathcal{M}_i \subset \mathcal{X}$, $i = 1, 2$ denote the fault modes. We denote the fault signatures associated with $\lambda(t) = i$ by L_{i1} and L_{i2}. The fault modes together with the fault signatures can be used to model the effects of actuator faults, sensor faults and system faults on the dynamics of the system. For example, the effect of a fault in the i-th actuator may be represented by L_{i1} as the i-th column of B_i and if an actuator fails, then $m_1(t) = -u_i(t)$.

The FPRG problem is concerned with the design of a Markovian jump residual generator that is governed by the filter dynamics of the form

$$\begin{aligned} \dot{w}(t) &= F(\lambda(t))w(t) - E(\lambda(t))y(t) + K(\lambda(t))u(t) \\ r(t) &= M(\lambda(t))w(t) - H(\lambda(t))y(t) \end{aligned} \tag{5.67}$$

where $w(t) \in \mathcal{F} \subset \mathcal{X}$ such that the response of $r(t)$ is affected by the fault mode $m_1(t)$ and is decoupled from $m_2(t)$ and if m_1 is identically zero then

$$\lim_{t\to\infty} \mathbb{E}||r(t)||^2 = 0$$

for any input signal $u(t)$.

We can rewrite equations (5.66) and (5.67) as follows

$$\begin{aligned} \begin{bmatrix} \dot{x}(t) \\ \dot{w}(t) \end{bmatrix} &= \begin{bmatrix} A(\lambda(t)) & 0 \\ -E(\lambda(t))C(\lambda(t)) & F(\lambda(t)) \end{bmatrix} \begin{bmatrix} x(t) \\ w(t) \end{bmatrix} \\ &\quad + \begin{bmatrix} B(\lambda(t)) & L_2(\lambda(t)) \\ K(\lambda(t)) & 0 \end{bmatrix} \begin{bmatrix} u(t) \\ m_2(t) \end{bmatrix} + \begin{bmatrix} L_1(\lambda(t)) \\ 0 \end{bmatrix} m_1(t) \\ r(t) &= \begin{bmatrix} -H(\lambda(t))C(\lambda(t)) & M(\lambda(t)) \end{bmatrix} \begin{bmatrix} x(t) \\ w(t) \end{bmatrix} \end{aligned} \tag{5.68}$$

Define the extended space $\mathcal{X}^e = \mathcal{X} \oplus \mathcal{F}$ and $\mathcal{U}^e = \mathcal{U} \oplus \mathcal{M}_2$, so that equation (5.68) can be expressed as

$$\begin{aligned} \dot{x}^e(t) &= A^e(\lambda(t))x^e(t) + B^e(\lambda(t))u^e(t) + L_1^e(\lambda(t))m_1(t) \\ r(t) &= H^e(\lambda(t))x^e(t) \end{aligned} \tag{5.69}$$

with $x^e(t) \in \mathcal{X}^e$ and $u^e \in \mathcal{U}^e$. Similar to the input observability of discrete-time Markovian jump systems, the same notion for continuous-time Markovian jump systems is now defined below.

Definition 5.13 *The input signal $m_1(t)$ is called input observable for the Markovian jump system* (5.69) *if L_i^{e1}, $i \in \Psi$ is monic and the image of L_i^{e1}'s does not intersect with the unobservable subspace of system* (5.69).

Based on the above definition, the FPRG problem can now be formally stated as the problem of designing the MJS dynamical filter (5.67) such that

(a) r is decoupled from u^e, (5.70)

(b) m_1 is input observable in the augmented system (5.69), and (5.71)

(c) $\lim_{t\to 0} \mathbb{E}\{||r(t)||^2\} = 0, \quad$ for $m_1(t) = 0, \forall\ i_0 \in \Psi$ and $\forall x_0^e \in \mathcal{X}^e$. (5.72)

It is easy to show that Lemma 5.7 holds also for the continuous-time MJS with the embedding map that is defined in (5.40). We are now in a position to derive the solvability condition for the FPRG problem corresponding to the Markovian jump system (5.66).

Theorem 5.10 *The FPRG problem has a solution for the augmented MJS system* (5.69) *only if*

$$\mathcal{S}^* \bigcap \mathcal{L}_{j1} = 0, \quad j \in \Psi \tag{5.73}$$

where $\mathcal{S}^* = \inf \mathfrak{S}(\mathfrak{A}, \mathfrak{C}, \sum_{i=1}^{N} \mathcal{L}_{i2})$. *On the other hand, if the above* $\mathcal{S}^*$ *exists such that it is also outer MS-detectable, then the FPRG problem is then guaranteed to have a solution.*

Proof: The proof is similar to the proof of Theorem 5.5 and is therefore omitted. ■

The next example illustrates through a detailed derivation steps how to design the detection filter (5.67) for a given continuous-time Markovian jump system.

Example 5.3. Consider the Markovian jump system (5.66) with $N = 2$ and matrices $A_i, C_i,\ i = 1, 2$ as follows

$$A_1 = \begin{bmatrix} 0 & 3 & 4 \\ 1 & 2 & 3 \\ 0 & 2 & 5 \end{bmatrix}, \quad A_2 = \begin{bmatrix} 1 & 2 & 4 \\ 2 & -1 & 2 \\ 0 & 1 & 4 \end{bmatrix}, C_1 = \begin{bmatrix} 0 & 1 & 0 \\ 0 & 0 & 1 \end{bmatrix}, C_2 = \begin{bmatrix} 1 & 1 & 0 \\ 0 & 0 & 1 \end{bmatrix}$$

and $B_i = 0$. The mode transition matrix and the fault signatures are given by

$$\Pi = \begin{bmatrix} -8 & 8 \\ 3 & -3 \end{bmatrix}, L_{i1} = \begin{bmatrix} 1 \\ 0 \\ 0 \end{bmatrix}, L_{i2} = \begin{bmatrix} -3 \\ 1 \\ 0 \end{bmatrix}, \quad i \in \Psi$$

respectively. The unobservability subspace $\mathcal{S}^*(\mathcal{L}_{i2})$ is obtained from Algorithm 5.3 which is given by

$$\mathcal{S}^*(\mathcal{L}_{i2}) = \begin{bmatrix} -3 \\ 1 \\ 0 \end{bmatrix}$$

It follows that $\mathcal{S}^*(\mathcal{L}_{i2}) \cap \mathcal{L}_{i1} = 0$. It can be checked that $\mathcal{S}^*(\mathcal{L}_{i2})$ is outer MS-detectable, and hence the FPRG problem has a solution. According to Theorem 5.10, the matrices that specify the governing dynamics of the detection filter in (5.67) are found through the following steps (all the geometric manipulations are performed by using the "geometric approach toolbox" [117]):

1. The output injection maps $D_i, i = 1, 2$ are obtained from $\mathcal{S}^*(\mathcal{L}_{i2})$ as

$$D_1 = \begin{bmatrix} 0 & -0.65 \\ 0 & -1.95 \\ -2 & -2.5 \end{bmatrix}, D_2 = \begin{bmatrix} -1.1 & -0.5 \\ -3.3 & -1.5 \\ 0.5 & -2.0 \end{bmatrix}$$

2. The measurement maps H_i are found from equation (5.25) as $H_1 = H_2 = \begin{bmatrix} 0 & 1 \end{bmatrix}$

3. The canonical projection map P for $\mathcal{S}^*(\mathcal{L}_{i2})$ is given by

$$P = \begin{bmatrix} 1 & -3 & 0 \\ 0 & 0 & 1 \end{bmatrix}$$

4. The maps $M_1 = M_2 = \begin{bmatrix} 0 & 1 \end{bmatrix}$ are the unique solutions to $M_i P = H_i C_i$, $i = 1, 2$
5. The induced maps A_{0i} are found from (2.23) as follows

$$A_{01} = \begin{bmatrix} 3 & -2.055 \\ 0 & 2.5 \end{bmatrix}, A_{02} = \begin{bmatrix} -4 & -1.581 \\ -1.581 & 2 \end{bmatrix}$$

6. The maps G_i are obtained by solving the corresponding LMI's by using the YALMIP LMI Toolbox [111] which yields

$$G_1 = \begin{bmatrix} 4.12 \\ -3.29 \end{bmatrix}, G_2 = \begin{bmatrix} 2.77 \\ -4.99 \end{bmatrix}$$

7. Finally, the maps F_i and E_i are found according to Theorem 5.10 as follows

$$F_1 = \begin{bmatrix} 3 & -2.0716 \\ 0 & -0.798 \end{bmatrix}, F_2 = \begin{bmatrix} -4 & 1.1941 \\ -1.581 & -2.99 \end{bmatrix},$$
$$E_1 = \begin{bmatrix} 0 & 6.18 \\ -2 & -5.798 \end{bmatrix}, E_2 = \begin{bmatrix} 3.47 & 4.35 \\ 0.5 & -6.99 \end{bmatrix}$$

It is interesting to note that the FPRG problem for the mode 1 by itself does not have a solution since $\mathcal{L}_{i1} \subset \mathcal{S}_1^*$. However, when there is a jump in the system, one can solve the FPRG problem. Figure 5.5 shows the residual signal $r(t)$ that is obtained for the above system by constructing the detection filter (5.67). Two fault scenarios are considered: (a) $m_2(t) = 0.1$, for $t > 10$ seconds, and (b) $m_1(t) = 0.1$, for $t > 15$ seconds. As shown in Figure 5.5, the residual signal is only affected by $m_1(t)$ and the fault mode $m_2(t)$ does not have any effect on it.

To conclude this section, we now consider a Markovian jump system that has multiple faults and that is governed by the following dynamical system

$$\begin{aligned} \dot{x}(t) &= A(\lambda(t))x(t) + B(\lambda(t))u(t) + \sum_{j=1}^{k} L_j(\lambda(t))m_j(t) \\ y(t) &= C(\lambda(t))x(t) \qquad x(0) = x_0, \quad \lambda(0) = i_0 \end{aligned} \tag{5.74}$$

where all the matrices are the same as in the dynamical model (5.66), $L_j(\lambda(k))$, $j \in \mathbf{k}$ denote the fault signatures, and $m_j(t) \in \mathcal{M}_j$, $j \in \mathbf{k}$ denote the fault modes. We denote the fault signatures associated with $\lambda(t) = i$ by $L_{ij}, i \in \Psi, j \in \mathbf{k}$.

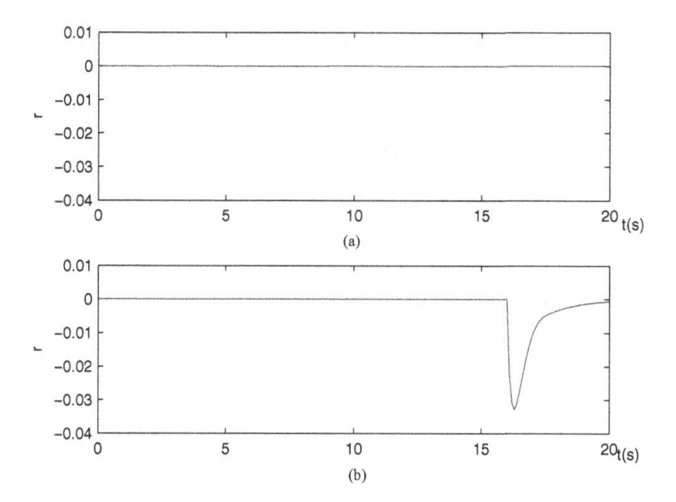

Fig. 5.5 The residual signals: (a) $m_2(t) = 0.1$, for $t > 10sec$; (b) $m_1(t) = 0.1$, for $t > 15sec$.

The SFDIP problem for the Markovian jump system (5.74) is now defined as the problem of generating p residual signals $r_j(t), j \in \mathbf{p}$ from the following Markovian jump detection filters

$$\begin{aligned} \dot{w}_j(t) &= F_j(\lambda(t))w_j(t) - E_j(\lambda(t))y(t) + K_j(\lambda(t))u(t) \\ r_j(t) &= M_j(\lambda(t))w_j(t) - H_j(\lambda(t))y(t), \qquad j \in \mathbf{k} \end{aligned} \tag{5.75}$$

such that the residuals $r_j(t)$ for $j \in \Omega_i$ are sensitive to a fault of the i-th component, and the other residuals $r_\alpha(t)$ for $\alpha \in \mathbf{p} - \Omega_i$ are insensitive to this fault. The solvability condition for the SFDIP problem is obtained by invoking the solvability condition that was developed earlier for the FPRG problem as follows.

Theorem 5.11 *The SFDIP problem has a solution for system* (5.74) *only if*

$$\mathcal{S}^*_{\Gamma_j} \bigcap \mathcal{L}^l_i = 0, \quad i \in \Psi, l \in \Gamma_j \tag{5.76}$$

where

$$\mathcal{S}^*_{\Gamma_j} = \inf \mathfrak{S}(\mathfrak{A}, \mathfrak{C}, \sum_{v=1}^{N} \sum_{l \notin \Gamma_j} \mathcal{L}^l_v), \quad j \in \mathbf{p} \tag{5.77}$$

On the other hand, if the above $\mathcal{S}^*_j, j \in \boldsymbol{k}$ exist such that they are outer MS-detectable, the EFPRG problem is then guaranteed to have a solution.

Proof: The proof is immediate by following along the same lines that were invoked for the proof of Theorems 5.10 and 2.3.

5.5.3 An H_∞-based Fault Detection and Isolation Design Strategy

In this section, we consider Markovian jump systems that are subjected to both external input and output measurement disturbances and noise and that are governed by

$$\begin{aligned}\dot{x}(t) &= A(\lambda(t))x(t) + B(\lambda(t))u(t) + \sum_{j=1}^{k} L_j(\lambda(t))m_j(t) + B_d(\lambda(t))d(t) \\ y(t) &= C(\lambda(t))x(t) + D_d(\lambda(t))d(t)\end{aligned} \tag{5.78}$$

where all the matrices are defined as in (5.74). The signal $d(t) \in \mathbb{R}^p$ represents an unknown disturbance input and output measurement noise. We denote the disturbance matrices $B_d(\lambda(t))$ and $D_d(\lambda(t))$ associated with $\lambda(t) = i$ by B_{di} and D_{di}, $i \in \Psi$. It is further assumed that the disturbance input $d(t)$ belongs to $\mathscr{L}_2[0, \infty]$, i.e.

$$||d(t)||_2 = \int_0^\infty d^\top(t)d(t)dt < \infty$$

Moreover, it is assumed that the mode transition matrix Π is not known precisely. In other words, it belongs to the following admissible uncertainty domain [198]

$$\mathscr{D}_\Pi = \{\bar{\Pi} + \Delta\Pi : |\Delta\Pi_{ij}| \leq \epsilon_{ij}, \epsilon_{ij} \geq 0, \;\; \forall i, j \in \Psi, \; i \neq j\} \tag{5.79}$$

where $\bar{\Pi} = [\bar{\pi}_{ij}]$ is a known constant matrix and denotes the estimated value of Π and $\Delta\Pi = [\Delta\pi_{ij}]$ denotes the uncertainty in the mode transition rate matrix.

According to the above representation of the Markovian jump system (5.78), an H_∞-based Extended Fundamental Problem in Residual Generation (HEFPRG) is formulated which is concerned with the design of a set of detection filters (5.75) that generate k residuals $r_j(t)$ such that a fault in the l-th component $m_l(t) \neq 0$ can only affect the residual $r_l(t)$ and no other residual $r_j(t) (j \neq l)$ and

$$||r_l||_{2,E}^2 = \mathbb{E}\Big\{ \int_0^\infty r_l^\top(t)r_l(t)dt|(x_0, i_0)\Big\} < \gamma^2 ||d||_2, \;\; l \in \mathbf{k} \tag{5.80}$$

for all $d(t) \in \mathscr{L}_2$ and $\Pi \in \mathscr{D}_\Pi$, where $\gamma > 0$ is the prescribed level of disturbance attenuation.

Below, we first present a preliminary result on the disturbance attenuation of continuous-time Markovian jump systems.

Lemma 5.10 *[24] Let γ be a given positive constant. If there exists a set of symmetric and positive-definite matrices $\mathfrak{R} = (R_1, ..., R_N) > 0$ such that the following set of coupled LMIs hold for every $i \in \Psi$:*

$$\begin{bmatrix} J_i & C_i^\top D_{di} + R_i B_{di} \\ * & D_{di}^\top D_{di} - \gamma^2 I \end{bmatrix} < 0 \tag{5.81}$$

where $J_i = A_i^\top R_i + R_i A_i + \sum_{j=1}^N \pi_{ij} R_j + C_i^\top C_i$, then system (5.78) with $u(t) = 0$ and $m_j(t) = 0, j \in \boldsymbol{k}$ is stochastically stable and for zero initial conditions satisfies the inequality

$$||y||_{2,E}^2 = \mathbb{E}\Big\{ \int_0^\infty y^\top(t) y(t) dt | (x_0, i_0) \Big\} < \gamma^2 ||d||_2, \quad l \in \boldsymbol{k} \tag{5.82}$$

A system that satisfies the above conditions is said to be stochastically stable with a γ-disturbance attenuation.

In the next lemma we consider the effects of uncertainties in the mode transition matrix Π for analyzing the stochastic stability properties of the system (5.78).

Lemma 5.11 *Let γ be a given positive constant. If there exists a set of symmetric and positive-definite matrices $\mathfrak{R} = (R_1, ..., R_N) > 0$ and $\{\xi_{ij} > 0, i, j \in \Psi, i \neq j\}$ such that the following set of coupled LMIs holds for every $i \in \Psi$:*

$$\begin{bmatrix} Q_i & C_i^\top D_{di} + R_i B_{di} & M_i \\ * & D_{di}^\top D_{di} - \gamma^2 I & 0 \\ * & * & -\Xi_i \end{bmatrix} < 0 \tag{5.83}$$

where

$$\begin{aligned} Q_i &= A_i^\top R_i + R_i A_i + \sum_{j=1}^N \bar{\pi}_{ij} R_j + C_i^\top C_i + \sum_{j=1, j\neq i}^N \frac{\xi_{ij}}{4} \epsilon_{ij}^2 I \\ M_i &= \begin{bmatrix} R_i - R_1 & \cdots & R_i - R_{i-1} & R_i - R_{i+1} & \cdots & R_i - R_N \end{bmatrix} \\ \Xi_i &= diag(\xi_{i1} I, ..., \xi_{i(i-1)} I, \xi_{i(i+1)} I, ..., \xi_{iN} I) \end{aligned}$$

then the uncertain system (5.78) with $u(t) = 0$ and $m_j(t) = 0, j \in \boldsymbol{k}$ is stochastically stable. Moreover, for zero initial conditions the system satisfies the inequality (5.82) for all $\Pi \in \mathscr{D}_\Pi$.

Proof: According to Lemma 5.10, the uncertain system (5.78) with $u(t) = 0$ and $m_j(t) = 0, j \in \mathbf{k}$ is stochastically stable with γ-disturbance attenuation if

$$A_i^\top R_i + R_i A_i + \sum_{j=1}^{N}((\bar{\pi}_{ij} + \Delta\pi_{ij})R_j + C_i^\top C_i - (C_i^\top D_{di} + R_i B_{di})(D_{di}^\top D_{di} - \gamma^2 I)^{-1}(C_i^\top D_{di} + R_i B_{di})^\top < 0$$

for all $i \in \Psi$. The above inequality can be rewritten as [198]

$$\begin{aligned} &A_i^\top R_i + R_i A_i + \sum_{j=1}^{N} \bar{\pi}_{ij} R_j + C_i^\top C_i \\ &+ \sum_{j=1, j\neq i}^{N} [\frac{1}{2}\Delta\pi_{ij}(R_j - R_i) + \frac{1}{2}\Delta\pi_{ij}(R_j - R_i)] \\ &- (C_i^\top D_{di} + R_i B_{di})(D_{di}^\top D_{di} - \gamma^2 I)^{-1}(C_i^\top D_{di} + R_i B_{di})^\top < 0 \end{aligned}$$

The above inequality holds for all $|\Delta\pi_{ij}| \leq \epsilon_{ij}$ if there exist $\xi_{ij} > 0, i, j \in \Psi, i \neq j$ such that

$$\begin{aligned} &A_i^\top R_i + R_i A_i + \sum_{j=1}^{N} \bar{\pi}_{ij} R_j + C_i^\top C_i + \sum_{j=1, j\neq i}^{N} [\frac{\xi_{ij}}{4}\epsilon_{ij}^2 I + \frac{1}{\xi_{ij}}(R_j - R_i)^2] \\ &- (C_i^\top D_{di} + R_i B_{di})(D_{di}^\top D_{di} - \gamma^2 I)^{-1}(C_i^\top D_{di} + R_i B_{di})^\top < 0 \end{aligned}$$

It can be shown easily that the above is equivalent to the inequality (5.83) by using the Schur complement. ■

We are now in a position to derive the sufficient conditions for determining the solvability of the HEFPRG problem for an uncertain MJS system.

Theorem 5.12 *The HEFPRG problem has a solution for the Markovian jump system (5.78) with uncertain mode transition matrix if there exist k outer MS-detectable unobservability subspaces*

$$\mathcal{S}_j^* = \inf \mathfrak{S}(\mathfrak{A}, \mathfrak{C}, \sum_{v=1}^{N} \sum_{l=1, l\neq j}^{k} \mathcal{L}_{vl}), \quad j \in \boldsymbol{k}$$

such that

$$\mathcal{S}_j^* \bigcap \mathcal{L}_{ij} = 0, \quad i \in \Psi, j \in \boldsymbol{k} \tag{5.84}$$

as well as the matrices T_{ij}, positive-definite matrices R_{ij}, $i \in \Psi$, $j \in \boldsymbol{k}$, and $\{\xi_{il}^j > 0, i, l \in \Psi, i \neq l, j \in \boldsymbol{k}\}$ such that

$$\begin{bmatrix} Q_{ij} & -M_{ij}^\top H_{ij} D_{di} - R_{ij} P_j D_{ij} D_{di} - R_{ij} P_j B_{di} - T_{ij} H_{ij} D_{di} & \Theta_{ij} \\ * & D_{di}^\top H_{ij}^\top H_{ij} D_{di} - \gamma^2 I & 0 \\ * & * & -\Xi_{ij} \end{bmatrix} < 0 \tag{5.85}$$

for all $i \in \Psi, j \in \boldsymbol{k}$ with

$$\begin{aligned} Q_{ij} =& A_{ij}^\top R_{ij} + M_{ij}^\top T_{ij}^\top + R_{ij} A_{ij} + T_{ij} M_{ij} + \sum_{l=1}^{N} \bar{\pi}_{il} R_{lj} + M_{ij}^\top M_{ij} \\ &+ \sum_{l=1, l \neq i}^{N} \frac{\xi_{il}^j}{4} \epsilon_{il}^2 I \\ \Theta_{ij} =& \left[R_{ij} - R_{1j} \cdots R_{ij} - R_{(i-1)j} \; R_{ij} - R_{(i+1)j} \cdots R_{ij} - R_{Nj} \right] \\ \Xi_{ij} =& diag(\xi_{i1}^j I, ..., \xi_{i(i-1)}^j I, \xi_{i(i+1)}^j I, ..., \xi_{iN}^j I) \end{aligned}$$

and where P_j is the canonical projection of $\mathcal{X}$ on $\mathcal{X}/\mathcal{S}_j^*$, the pairs (M_{ij}, A_{ij}), $i \in \Psi$, $j \in \boldsymbol{k}$ are the factor system of the pairs $(C_i, A_i), i \in \Psi$ on $\mathcal{X}/\mathcal{S}_j^*$, H_{ij} is the solution to $Ker\ H_{ij} C_i = \mathcal{S}_j^* + Ker\ C_i$ and $\mathcal{S}_j^* =<< \bigcap_{l=1}^{N} Ker\ H_{lj} C_l | A_i + D_{ij} C_i >>_{i \in \Psi}$, $j \in \boldsymbol{k}$.

Proof: Given P_j as the canonical projection of $\mathcal{X}$ on $\mathcal{X}/\mathcal{S}_j^*$, let $M_{ij}, i \in \Psi$ denote a unique solution to $M_{ij} P_j = H_{ij} C_i$ and define $A_{ij} = (A_i + D_{ij} C_i : \mathcal{X}/\mathcal{S}_j^*), i \in \Psi$. Let $G_{ij} = R_{ij}^{-1} T_{ij}$, $i \in \Psi, j \in \mathbf{k}$ where T_{ij} and R_{ij} are solutions to the inequality (5.85). Define $F_{ij} = A_{ij} + G_{ij} M_{ij}$, $E_{ij} = P_j (D_{ij} + P_j^{-r} G_{ij} H_{ij})$ and $K_{ij} = P_j B_i$ for $i \in \Psi, j \in \mathbf{k}$. Furthermore, define $e_j(t) = w_j(t) - P_j x(t)$, so that by using (5.75) we have

$$\begin{aligned} \dot{e}_j(t) =& F_{ij} w_j(t) - E_{ij} y(t) + K_{ij} u(t) \\ &- P_j (A_i x(k) + B_i u(k) + \sum_{l=1}^{k} L_{il} m_l(t) + B_{di} d(t)) \\ =& (A_{ij} + G_{ij} M_{ij}) e_j(t) - P_j L_{ij} m_j(t) \\ &- P_j D_{ij} D_{di} d(t) - P_j B_{di} d(t) - G_{ij} H_{ij} D_{di} d(t) \end{aligned}$$

Also

$$\begin{aligned} r_j(t) &= M_{ij} w_j(t) - H_{ij} y(t) = M_{ij} w_j(t) - H_{ij} C_i x(t) - H_{ij} D_{di} d(t) \\ &= M_i e_j(t) - H_{ij} D_{di} d(t) \end{aligned}$$

Consequently, the error dynamics associated with the detection filters can be expressed as

$$\begin{aligned}\dot{e}_j(t) =&(A_{ij} + G_{ij}M_{ij})e_j(t) - P_jL_{ij}m_j(t)\\ &- P_jD_{ij}D_{di}d(t) - P_jB_{di}d(t) - G_{ij}H_{ij}D_{di}d(t)\\ r_j(t) =&M_{ij}e_j(t) - H_{ij}D_{di}d(t)\end{aligned} \tag{5.86}$$

Therefore, the residual r_j is only affected by the fault m_j and according to Lemma 5.11 and inequality (5.85), the disturbance attenuation inequality (5.80) holds for all the residuals $r_j(t),\ j \in \mathbf{k}$. ■

Remark 5.4. It should be noted in the above theorem that since we only considered the sufficient solvability conditions for the HEFPRG problem, there is no need to consider an augmented system for the purpose of analysis and proof.

Once the residual signals $r_j(t), j \in \mathbf{k}$ are constructed and generated, the final step in developing a reliable fault detection and isolation strategy deals with the residual evaluation process which involves determining the evaluation functions J_{r_j} and their associated thresholds J_{th_j}. In this section, the evaluation functions and the thresholds are selected, respectively according to the following formal criteria, namely

$$J_{r_j}(t) = \int_{t-T_0}^{T} r_j^{\top}(\tau)r_j(\tau)d\tau, \quad j \in \mathbf{k} \tag{5.87}$$

$$J_{th_j} = \sup_{d\in\mathfrak{L}_2, m_j=0} \mathbb{E}(J_{r_j}), \quad j \in \mathbf{k} \tag{5.88}$$

where T_0 is the length of the evaluation window. According to the above thresholds and evaluation functions, the occurrence of a fault can be detected and isolated by using the following decision logics

$$J_{r_j} > J_{th_j} \Longrightarrow m_j \neq 0, \quad j \in \mathbf{k} \tag{5.89}$$

5.5.4 A Case Study

In this section, the proposed H_∞-based FDI algorithm is applied to a VTOL (vertical take-off and landing) helicopter [140, 24] model. The dynamics of the VTOL system can be written as

$$\begin{aligned}\dot{x}(t) &= A(\lambda(t))x(t) + B(\lambda(t))u(t) + L_1(\lambda(t))m_1(t) + L_2(\lambda(t))m_2(t) + B_dd(t)\\ y(t) &= Cx(t) + D_dd(t)\end{aligned} \tag{5.90}$$

where $\lambda(t)$ is a continuous time Markov process with three different modes that correspond to the airspeeds 135 (nominal value), 60 and 170 knots. The state variables corresponding to the system are taken as the horizontal velocity (x_1), the vertical velocity (x_2), the pitch rate (x_3), and the pitch angle (x_4). The input signals u_1 and u_2 are the collective pitch control and the longitudinal cyclic pitch control, respectively. The input signal $d(t)$ represents the external disturbances and uncertainties. The matrices associated with the VTOL system are given by

$$A(\lambda(t)) = \begin{bmatrix} -0.04 & 0.04 & 0.02 & -0.5 \\ 0.05 & -1.01 & 0.0 & -4.0 \\ 0.1 & a_{32}(\lambda(t)) & -0.71 & a_{34}(\lambda(t)) \\ 0.0 & 0.0 & 1.0 & 0.0 \end{bmatrix}, B(\lambda(t)) = \begin{bmatrix} 0.44 & 0.18 \\ b_{21}(\lambda(t)) & -7.6 \\ -5.52 & 4.49 \\ 0 & 0 \end{bmatrix}$$

$$C = \begin{bmatrix} 1\,0\,0\,0 \\ 0\,1\,0\,0 \\ 0\,0\,1\,0 \end{bmatrix}, B_d = \begin{bmatrix} 0.1\ 0.0 \\ 0.0\ 0.1 \\ 0.1\ 0.0 \\ 0.0\ 0.1 \end{bmatrix}, D_d = \begin{bmatrix} 0.0\ 0.1 \\ 0.1\ 0.1 \\ 0.0\ 0.0 \end{bmatrix}$$

where the values of the parameters $a_{32}(\lambda(t))$, $a_{34}(\lambda(t))$, and $b_{21}(\lambda(t))$ are given in Table 5.1. The fault signatures $L_1(\lambda(t))$ and $L_2(\lambda(t))$ represent the actuator faults, and hence are selected as the first and the second columns of $B(\lambda(t))$. Different fault modes can be considered for actuators in general [23], namely, i) lock-in-place (LIP), ii) loss of effectiveness, iii) float, and iv) hard over. For this case study we consider LIP, loss of effectiveness, and float faults. The mode transition matrix is taken as

$$\Pi = \begin{bmatrix} -2.09 & 1.07 & 1.02 \\ 0.07 & -0.07 & 0.0 \\ 0.02 & 0.0 & -0.02 \end{bmatrix}$$

with an uncertainty level of $\epsilon_{ij} = 0.1, i, j \in \Psi i \neq j$ as specified in equation (5.79).

Table 5.1 The specific parameter values of the VTOL helicopter

Airspeed (knots)	$a_{32}(\lambda(t))$	$a_{34}(\lambda(t))$	$b_{21}(\lambda(t))$
135	0.37	1.42	3.55
60	0.07	0.12	1.0
170	0.51	2.52	5.11

The FDI problem that is considered for the VTOL is to design two residual signals $r_1(t)$ and $r_2(t)$ such that $r_1(t)$ is only affected by the first actuator fault ($m_1(t)$) and $r_2(t)$ is only affected by the second actuator fault ($m_2(t)$). Moreover, the effects of the disturbance input $d(t)$ on these residuals are attenuated with a factor of γ. According to the results of Theorem 5.12, we first need to construct the unobservability subspaces $\mathcal{S}_1^* = \inf(\mathfrak{A}, \mathfrak{C}, \mathcal{L}_2)$ and $\mathcal{S}_2^* = \inf(\mathfrak{A}, \mathfrak{C}, \sum_{i=1}^{3} \mathcal{L}_{i1})$. These unobservability subspaces are obtained by using the Algorithm 5.3 as follows

$$\mathcal{S}_1^* = \begin{bmatrix} 0.0204 \\ -0.8608 \\ 0.5085 \\ 0 \end{bmatrix}, \mathcal{S}_2^* = \begin{bmatrix} -0.0795 & 0 \\ 0 & 1 \\ 0.9968 & 0 \\ 0 & 0 \end{bmatrix}$$

It can be easily verified that the necessary conditions $\mathcal{S}_1^* \cap \mathcal{L}_{i1} = 0, i = 1, 2, 3$ and $\mathcal{S}_2^* \cap \mathcal{L}_2 = 0$ are both satisfied. Next, we need to verify the feasibility of the inequality (5.85) corresponding to a given disturbance attenuation level γ. Using the LMI toolbox, inequality (5.85) is solved for $\gamma = 0.1$. Subsequently, all the maps and matrices that are defined in Theorem 5.12 are obtained. It should be noted that from equation (5.75), $r_1 \in \mathbb{R}^2$ and $r_2 \in \mathbb{R}$.

The disturbance inputs $d_1(t)$ and $d_2(t)$ are assumed to be independent and band-limited white-noises with the power of 0.1 and 0.2, respectively. The length of the evaluation window is selected as $T_0 = 2$ seconds. The calculated thresholds are found to be $J_{th_1} = J_{th_2} = 5e - 3$. Figure 5.6 shows both the residuals and the evaluation functions corresponding to a float fault that is injected in the first actuator (u_1) at $t = 25$ seconds (float fault implies that the actuator is frozen at zero output). This fault can be modeled as $m_1(t) = -u_1(t)$, where $m_1(t)$ is the fault mode of the first actuator. As shown in Figure 5.6, the fault is detected and isolated at $t = 34$ seconds and the evaluation function of r_2 (i.e. J_{r_2}) remains below its corresponding threshold. Figure 5.7 shows the residuals and the evaluation functions corresponding to an intermittent fault that is injected in the second actuator where the actuator is locked (LIP fault) at a value of 0.1 between $t = 20$ seconds and $t = 30$ seconds. This fault can be modeled as $m_2(t) = -u_2(t) + 0.1$, where $m_2(t)$ is the fault mode of the second actuator. As shown in Figure 5.7, this fault is detected and isolated at $t = 22.5$ seconds and the evaluation function of r_1 (i.e. J_{r_1}) remains below its associated threshold.

Figure 5.8 shows the residuals and the evaluation functions corresponding to simultaneous faults that are injected in both actuators where a 50% loss of effectiveness (gain) fault is occurred in the first actuator at $t = 25$ seconds and the second actuator is locked at the output value of 0.1 between $t = 20$ seconds and $t = 30$ seconds. According to Figure 5.8, both faults are properly detected and isolated at time $t = 25$ seconds and $t = 26$ seconds, respectively.

5.6 Conclusions

The problem of fault detection and isolation in a network of unmanned vehicles in the presence of imperfect communication links is investigated in this chapter in the framework of Markovian jump systems. A geometric approach to the problem of fault detection and isolation of both continuous-time and discrete-time linear Markovian jump systems is developed. Starting with a new geometric characterization of the unobservable subspace of a Markovian

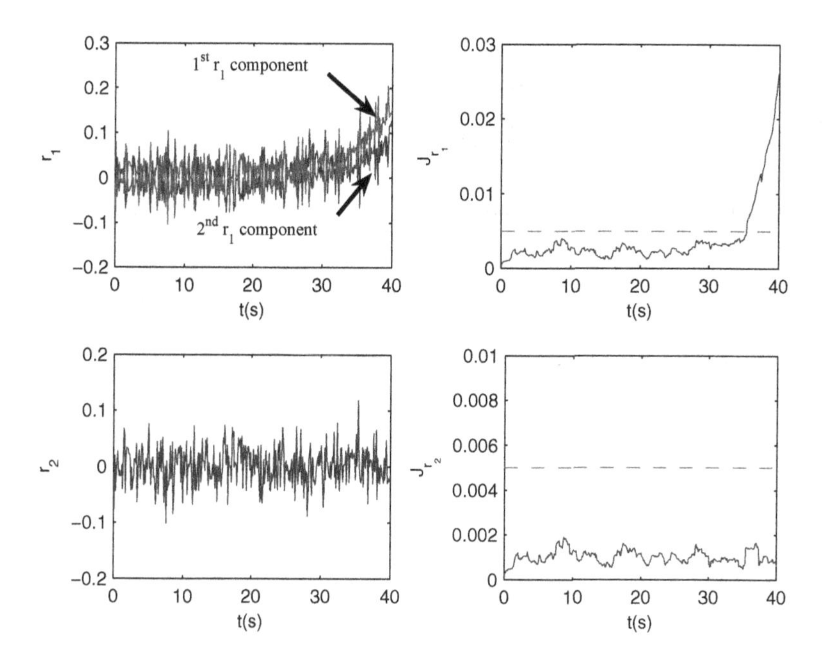

Fig. 5.6 The residual signals and their evaluation functions corresponding to a float fault in the first actuator $m_1 = -u_1$ (the dashed line in J_{r_i} denotes the threshold).

jump system, the concept of unobservability subspaces is formalized and an algorithm for constructing these subspaces is presented. By invoking the notion of unobservability subspace, the necessary and sufficient conditions for solving the fundamental problem of residual generation for Markovian jump systems is formally investigated. For uncertain Markovian jump systems, an H_∞-based fault detection and isolation strategy is proposed and developed where a set of residual signals are constructed such that each residual is only affected by one fault and is decoupled from the others while the H_∞ norm of the transfer function between the unknown input (external disturbances and output measurement noise) and the residual signals is guaranteed to be less than a prescribed desired value. Simulation results for application of the proposed novel methodologies to a VTOL (vertical take-off and landing) helicopter and network of formation flight of satellites are also presented to demonstrate and illustrate their effectiveness and capabilities.

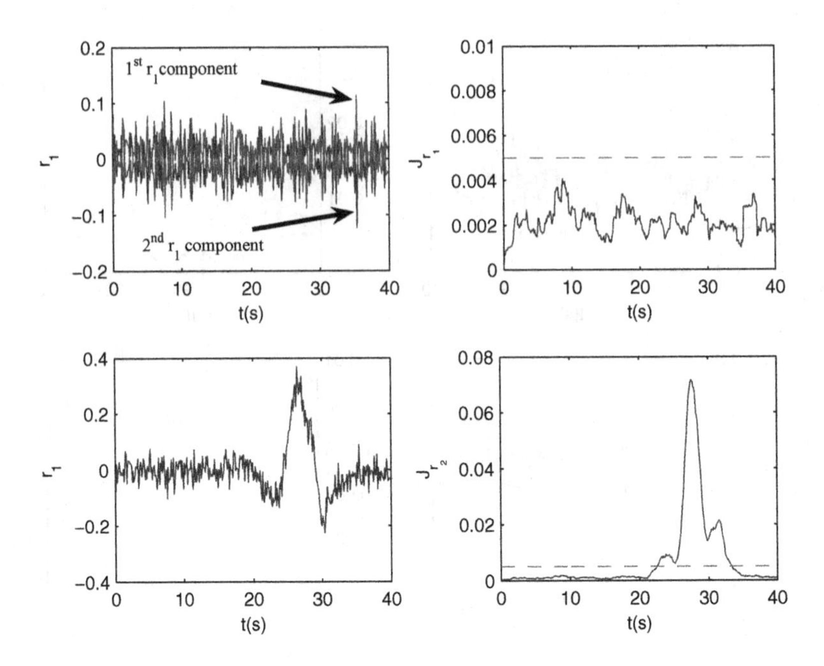

Fig. 5.7 The residual signals and their evaluation functions corresponding to a lock-in-place (LIP) fault in the second actuator $m_2 = -u_2 + 0.1$ (the dashed line in J_{r_i} denotes the threshold).

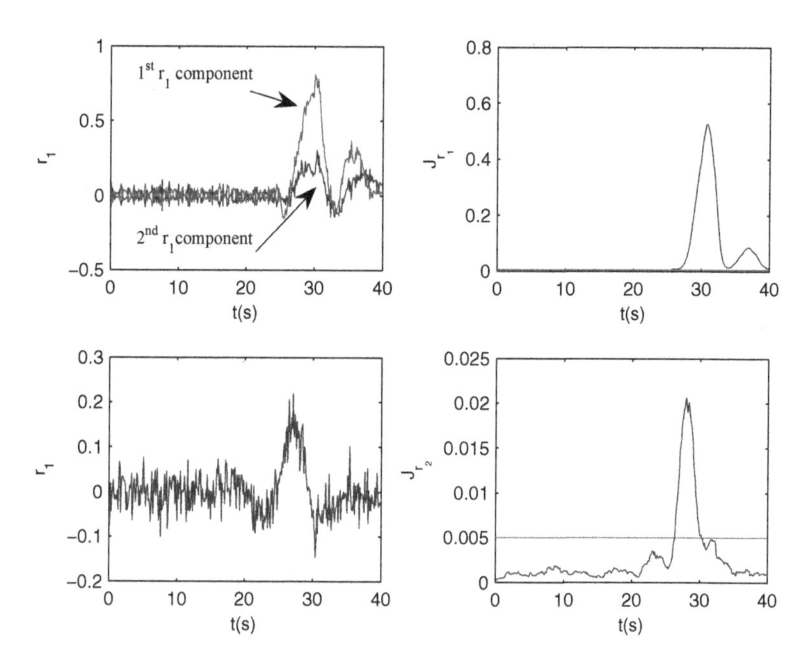

Fig. 5.8 The residual signals and their evaluation functions corresponding to multiple faults in actuators, namely a 50% loss of effectiveness fault in the first actuator and a lock-in-place fault in the second actuator (the dashed line in J_{r_i} denotes the threshold).

Chapter 6
Perspectives and Future Directions of Research

In this book, we have been interested in design and analysis of fault detection and isolation (FDI) strategies for networked multi-vehicle unmanned systems. This problem is timely and important due to the fact that one of the main challenges in these systems is developing an autonomous cooperative control solution that can maintain the group or team behavior and mission performance in presence of undesirable events such as faults in vehicles. In order to have an autonomous network of unmanned vehicles, fault detection and isolation schemes should be developed that are capable of detecting and isolating faults in the vehicles. The approach and framework that is proposed and presented here is based on geometric FDI. We have formulated and introduced several problems within the domain of multi-vehicle systems and obtained some very novel results.

In Chapters 3, 4, and 5, we have tackled various issues in the design of FDI algorithms for a network of unmanned vehicles. These include issues such as different FDI architectures, robustness with respect to external disturbances, and compensations for the effects of non-ideal communication links. Obviously, there is more work that needs to be done to have a complete theory on fault detection and isolation for networked multi-vehicle systems. In this chapter, a brief summary of each chapter is provided and a couple of major open problems in each area are identified.

6.1 Future Research Directions

In this section, we summarize the major research areas that are addressed in this book and identify some of the open problems in each area.

6.1.1 Fault Detection and Isolation of a Network of Unmanned Vehicles: Ideal Communication Channels

In Chapter 3, we dealt with the problem of fault detection and isolation for a network of unmanned vehicles without considering the effects of communication channels on the performance of the proposed FDI schemes. It was shown that actuator faults signatures in multi-vehicle systems with relative state measurements are dependent, and hence the entire network is overactuated. New coding schemes are developed for both linear and nonlinear systems with dependent fault signatures. Three case studies are considered as potential applications of these novel coding schemes. In the first case study, the FDI problem in a network of unmanned vehicles with relative state measurements is solved corresponding to three different architectures, namely, centralized, decentralized, and semi-decentralized architectures. It was shown that the solvability conditions for the centralized and semi-decentralized architectures are identical for vehicles with more than one neighbor. Moreover, it was shown that vehicles cannot perform fault detection and isolation in the fully decentralized architecture without an exchange of information among themselves. In the second case study, the actuator fault detection and isolation problem in an F18-HARV aircraft was presented as an example of overactuated linear systems. Finally, in the third case study, fault detection and isolation problem in a satellite with redundant reaction wheels is considered as an example of overactuated nonlinear systems.

The major directions for future research that can be pursued in this area are as follows:

- Development of FDI algorithms for a network of unmanned vehicles with both absolute (internal) and relative state measurements.
- Investigation of the effects of time-varying network topologies on the FDI schemes.
- Development of robust FDI algorithms with respect to modeling errors and unmodeled dynamics.

6.1.2 Fault Detection and Isolation of a Network of Unmanned Vehicles Subject to Large Environmental Disturbances

In Chapter 4, we proceeded by considering the effects of environmental disturbances on the vehicles and a hybrid fault detection and isolation scheme was developed for achieving robustness with respect to these disturbances. A hybrid architecture for a robust FDI is introduced that is composed of a bank of continuous-time residual generators and a DES fault diagnoser. It was shown that the proposed hybrid FDI algorithm is applicable to both linear and nonlinear systems.

The directions for future research that can be pursued in this area are as follows:

- Investigation of the effects of time-varying network topologies on the FDI schemes.
- Development of robust fault identification for a network of unmanned vehicles subject to external disturbances.

6.1.3 Fault Detection and Isolation of a Network of Unmanned Vehicles with Imperfect Communication Links

In Chapter 5, we considered a network of unmanned vehicles subject to non-ideal and imperfect communication channels. The packet erasure channel model was considered for communication links and the integration of this model with the vehicles dynamics yielded a discrete-time Markovian jump representation for the entire network. A geometric approach was developed for fault detection and isolation of both discrete-time and continuous-time Markovian jump systems. A notion of unobservability subspace was introduced for Markovian jump systems. The proposed FDI scheme was applied to the formation flight of satellites with imperfect communication links. Furthermore, to demonstrate the applicability of the proposed algorithms to also single vehicle systems, the actuator fault detection and isolation for an VTOL helicopter is considered as a case study for continuous-time Markovian jump systems.

The major directions for future research that can be pursued in this area are as follows:

- Development of the geometric property of weak controllability for Markovian jump systems and defining the notion of controllability subspaces for MJS.
- Development of FDI algorithms for discrete-time Markovian jump systems which are robust with respect to uncertainties in the transition matrix.
- Development of FDI algorithms for nonlinear Markovian jump systems. This problem has a potential application to a network of unmanned vehicles with nonlinear dynamics.
- Development of geometric framework for nonlinear Markovian jump systems and introducing the notions such as observability codistribution and controllability distribution.

References

1. K. Adjallah, D. Maquin, and J. Ragot. Nonlinear observer-based fault detection. *Proceeding of the third IEEE Conference on Control Application*, pages 1150–1120, 1994.
2. K. Adjallah, D. Maquin, and J. Ragot. Robust observer-based fault diagnosis in nonlinear uncertain systems. *Proceeding of the third IEEE Conference on Control Application*, pages 1150–1120, 1994.
3. I. M. Al-Salami, S. X. Ding, and P. Zhang. Statistical based residual evaluation for fault detection in networked control systems. *2006 IAR Annual meeting*, 2006.
4. I.A. Al-Zyoud and K. Khorasani. Detection of actuator faults using dynamics neural network for the attitude control subsystem of a satellite. *Proc. IEEE International Joint Conference on Neural Networks (IJCNN-2005)*, pages 1746–1751, 2005.
5. M. O. Anderson, M. D. Mckay, and B. S. Richardson. Multirobot automated indoor floor characterization team. *Proceeding of IEEE Conference on Robotics and Automation*, pages 1750–1753, 1996.
6. Anon. U.S. Military Specification MIL-F-8785C. 1980.
7. N. Aouf, B. Boulet, and R. Botez. H_2 and H_∞ optimal gust load alleviation for a flexible aircraft. *Proceeding of American Control Conference*, pages 1872–1876, 2000.
8. A. Barua, P. Sinha, K. Khorasani, and S. Tafazoli. A novel fault-tree approach for identifying potential causes of satellite reaction wheel failure. In *Proc. 2005 IEEE International Conference on Control Applications (CCA-2005)*, pages 1467–1472, Toronto, Canada, August 2005.
9. G. Basile and G. Marro. On the robust controlled invariant. *Systems & Control Letters*, 9(3):191–195, 1987.
10. G. Basile and G. Marro. *Controlled and conditioned invariant subspaces in linear system theory*. Prentice-Hall, 1992.
11. M. Basseville. Detecting changes in signals and systems - a survey. *Automatica*, 24(3):309–326, 1988.
12. M. Basseville and I. V. Nikiforov. *Detection of Abrupt changes: theory and application*. Information and System Science, Prentice-Hall, New York, 1993.
13. R. V. Beard. *Failure accommodation in linear system through self reorganization*. PhD thesis, Massachusettes Inst. Technol, 1971.
14. R. W. Beard, J. Lawton, and F.Y. Hadaegh. A coordination architecture for spacecraft formation control. *IEEE Trans. Control Syst. Technol.*, 9(6):777–790, 2001.

15. R. W. Beard and T. W. McLain. Multiple UAV cooperative search under collision avoidance and limited range communication constraints. *IEEE Conference on Decision and Control*, pages 25–30, 2003.
16. Randal W. Beard, Timothy W. McLain, Michael Goodrich, and Erik P. Anderson. Coordinated target assignment and intercept for unmanned air vehicles. *IEEE Transactions on Robotics and Automation*, 18(6):911–922, 2002.
17. A. Benallegue, A. Mokhtari, and L. Fridman. High-order sliding-mode observer for a quadrotor UAV. *International journal of robust and nonlinear control*, 18:427–440, 2008.
18. J. G. Bender. An overview of systems studies of automated highway systems. *IEEE Transactions on Vehicular Technology*, 40(1):82–99, 1991.
19. D. Berdjag, C. Christophe, and V.t Cocquempot. Nonlinear model decomposition for fault detection and isolation system design. *Proceedings of 45th IEEE Conference on Decision and Control*, pages 3321–3326, 2006.
20. L. Berec. A multi-model method to fault detection and diagnosis: Bayesian solution. an introductory treatise. *Int. J. of Adaptive Contr. and Signal Processing*, 44(6):1009–1016, 1998.
21. P. Bhatta, E. Fiorelli, F. Lekien, N. E. Leonard, D. A. Paley, F. Zhang, R. Bachmayer, R. E. Davis, D.M. Fratantoni, and R. Sepulchre. Coordination of an underwater glider fleet for adaptive ocean sampling. *Proc. Int. Workshop on Underwater Robotics for Sustainable Management of Marine Ecosystems and Environmental Monitoring*, pages 61–69, 2005.
22. J. D. Boskovic, S. E. Bergstrom, and R. K. Mehra. Retrofit reconfigurable flight control in the presence of control effector damage. *Proceeding of American Control Conference*, pages 2652–2657, 2005.
23. J. D. Boskovic, S. E. Bergstrom, and R. K. Mehra. Retrofit reconfigurable flight control in the presence of control effector damage. *Proceedings of the 2005 American Control conference*, pages 2652–2657, 2005.
24. El-Kebir Boukas. *Stochastic Switching Systems: Analysis and Design.* Birkhauser, Boston, 2006.
25. L. E. Buzogany, M. Pachter, and J. J. d'Azzo. Automated control of aircraft in formation flight. *Proceedings of the AIAA Conference on Guidance, Navigation, and Control*, pages 1349–1370, 1993.
26. Y. Cao and J. Lam. Robust H_∞ control of uncertain Markovian jump systems with time-delay. *IEEE Trans. Autom. Control*, 45(1):77–83, 2000.
27. A. Casavola, D. Famularo, and G. Franze. A robust deconvolution scheme for fault detection and isolation of uncertain linear systems: an LMI approach. *Automatica*, 41:1463–1472, 2005.
28. D. W. Casbeer, D. B. Kingston, R. W. Beard, S. Li T. W. McLain, and R. Mehra. Cooperative forest fire surveillance using a team of small unmanned air vehicles. *International Journal of Systems Sciences*, 37(6):351–360, 2006.
29. P. Castillo, A. Dzul, and R. Lozano. Real-time stabilization and tracking of a four-rotor mini rotorcraft. *IEEE Transactions on Control System Technology*, 12(4):510–516, 2004.
30. J. Chen. *Robust residual generation for model-based fault diagnosis of dynamic systems.* PhD thesis, University of York, York, UK, 1995.
31. J. Chen and R. J. Patton. *Robust model-based fault diagnosis for dynamic systems.* Kluwer Academic Publishers, Boston/Dordrecht/London, 1999.
32. J. Chen and H. Y. Zhang. Parity vector approach for detecting failures in dynamic systems. *Int. J. Sys. Sci.*, 21(4):765–770, 1990.
33. R. H. Chen, D. L. Mingori, and J. L. Speyer. Optimal stochastic fault detection filter. *Automatica*, 39(3):377–390, 2003.
34. R. H. Chen and J. L. Speyer. A generalized least-squares fault detection filter. *International Journal of Adaptive Control and Signal Processing*, 14:747–757, 2000.

35. R. H. Chen and J. L. Speyer. Robust multiple-fault detection filter. *International Journal of Robust and Nonlinear Control*, 12(8):675 – 696, 2002.
36. W. Chen and M. Saif. Fault detection and isolation based on novel unknown input observer design. *Proceedings of the 2006 American Control Conference*, pages 5129–5134, 2006.
37. W. Chen and M. Saif. A sliding mode observer-based strategy for fault detection, isolation, and estimation in a class of lipschitz nonlinear systems. *International Journal of Systems Science*, 38(12):943–955, 2007.
38. Z. Chen, Y. He, F. Chu, and J. Huang. Evolutionary strategy for classification problems and its application in fault diagnosis. *Engineering Applications of Artificial intelligence*, 16(1):31–38, 2003.
39. L. Cheng and Y. Wang. Fault tolerance for communication-based multirobot formation. *Proceedings of the Third International Conference on Machine Leming and Cybernetics*, pages 127–132, 2004.
40. D. F. Chichka, J. L. Speyer, and C. G. Park. Peak-seeking control with application to formation flight. *IEEE Conference on Decision and Control*, pages 2463–2470, 1999.
41. E. Y. Chow and A. S. Willsky. Analytical redundancy and the design of robust detection systems. *IEEE Trans. Autom. Control*, 29(7):603–614, 1984.
42. W. H Chung and J. L. Speyer. A game theoretic fault detection filter. *IEEE Transaction on Automatic Control*, 43(2):143–161, 1998.
43. W. H Chung and J. L. Speyer. A decectralized fault detection filter. *ASME J. of Dynamic Systems, Measurement, Control*, 123(2):237–247, 2001.
44. R. N. Clark. A simplified instrument failure detection scheme. *IEEE Trans. Aerosp. Electron. Syst.*, 14(4):558–563, 1978.
45. E. F. Costa and J. B. R. Do Val. On the observability and detectability of continuous-time Markov jump linear systems. *Siam J. Control Optim.*, 41(4):1295–1314, 2002.
46. E. F. Costa and J. B. R. Do Val. Weak detectabilty and the linear-quadratic control problem of discrete-time Markov jump linear systems. *International Journal of Control*, 75(16/17):1282–1292, 2002.
47. O. L. V. Costa, M. D. Fragoso, and R. P. Marques. *Jump linear systems in automatic control.* Marcel Dekker Inc., New York, 1990.
48. O. L. V. Costa, M. D. Fragoso, and R. P. Marques. *Discrete-Time Markov Jump systems.* Springer-Verlag, London, 2005.
49. M. Daigle, X. Koutsoukos, and G. Biswas. Fault diagnosis of continuous systems using discrete-event methods. *Proceedings of 46th IEEE Conference on Decision and Control*, 2007.
50. M. J. Daigle, X. D. Koutsoukos, and G. Biswas. Distributed diagnosis in formations of mobile robots. *IEEE Transaction on Robotics*, 23(2):353–369, 2007.
51. M. C. de Oliveira, J. C. Geromel, and J. Bernussou. Extend H_2 and H_∞ norm characterizations and controller parameterizations for discrete time systems. *International Journal of Control*, 75(9):666–679, 2002.
52. C. E. de Souza and M. D. Fragoso. H_∞ filtering for Markovian jump linear systems. *International Journal of Systen Science*, 33:909–915, 2002.
53. J. P. Desai, J. Ostrowski, and V Kumar. Controlling formations of multiple mobile robots. *IEEE International Conference on Robotics and Automation*, 4:2864 – 2869, 1998.
54. W. C. Dickson, R. H. Cannon, and S. M. Rock. Symbolic dynamic modeling and analysis of object/robot-team systems with experiment. *Proceeding of IEEE Conference on Robotics and Automation*, pages 1413–1420, 1996.
55. S. X. Ding, T. Jeinsch, P. M. Frank, and E. L. Ding. A unified approach to the optimization of fault detection system. *International Journal of Adaptive Control and Signal Processing*, 14(7):725–745, 2000.

56. S. X. Ding, P. Zhang, P. M. Frank, and E. L. Ding. Threshold calculation using LMI-technique and its integration in the design of fault detection systems. *Proceeding of the 42th Conference on Decision and Control*, pages 469–474, 2003.
57. X. Ding and L. Guo. An approach to time domain optimization of observer-based fault detection systems. *International Journal of Control*, 69(3):419–442, 1998.
58. R. K. Douglas and J. L. Speyer. Robust detection filter design. *Proceedings of the American Control Conference*, pages 91–96, 1995.
59. R. K. Douglas and J. L. Speyer. h_∞ bounded detection filter. *AIAA Journal of Guidance, Control, and Dynamics*, 22(1):129–138, 1999.
60. E. Earon. Almost-lighter-than-air-vehicle fleet simulation. *Technical Report, V. 0.9, Quanser Inc., Toronto, Canada*, 2005.
61. C. Edwards, S. K. Spurgeon, and R. J. Patton. Sliding mode observers for fault detection and isolation. *Automatica*, 36(4):541–553, 2000.
62. E. O. Elliott. Estimates of error rates for codes on burst-noise channels. *Bell Systems Technical Journal*, 42, 1963.
63. A. Fax and R. Murray. Graph laplacians and vehicle formation stabilization. *Proceedings of the 2002 IFAC World Congress*, 2002.
64. A. Fax and R. Murray. Information flow and cooperative control of vehicle formations. *Proceedings of the 2002 IFAC World Congress*, 2002.
65. P. M. Frank. Fault diagnosis in dynamic system via state estimation- a survey. In S. G. Tzafestas, M. G. Singh, and G. Schmidts, editors, *System Fault Diagnostics, Reliability and Related Knowledge-based Approaches*, volume 1, pages 35–98. D. Reidel Press, Dordrecht, 1987.
66. P. M. Frank. Fault diagnosis in dynamics system using analytical and knowledge-based redundancy- a survey and some new results. *Automatica*, 26(3):459–474, 1990.
67. P. M. Frank and X. Ding. Frequency domain approach to optimally robust residual generation and evaluation for model-based fault diagnosis. *Automatica*, 30(4):786–804, 1994.
68. C. V. M. Fridlund. Darwin - the infrared space interferometry mission. *ESA Bulletin 103*, 2000.
69. M. J. Fuente and S. Saludes. Fault detection and isolation in a nonlinear plant via neural networks. *Proceeding of the 4th IFAC Symp. Fault Detection, Supervision and Safety of Technical Processes, SAFEPROCESS'2000*, 1:472–477, 2000.
70. E. A. Garci and P. M. Frank. Deterministic nonlinear observer-based approaches to fault diagnosis: A survey. *IFAC Control Eng. Prac.*, 5:663–670, 1997.
71. J. P. Gauthier and I. A. K. Kupca. Observability and observers for nonlinear systems. *SIAM Journal on Control and Optimization*, 32:975–994, 1994.
72. J. Gertler. Survey of model-based failure detection and isolation in complex plants. *IEEE Control Syst. Mag.*, 8(6):3–11, 1988.
73. J. Gertler. Analytical redundacy methods in failure detection and isolation. *Proceeding of IFAC Fault Detection, Supervision and Safety for Technical Processes*, pages 9–21, 1991.
74. J. Gertler. Fault detection and isolation using parity relations. *Control Engineering Practice*, 5(5):653–661, 1997.
75. J. Gertler and D. Singer. A new structural framework for parity equation based failure detection and isolation. *Automatica*, 26:381–388, 1990.
76. J. J. Gertler. *Fault Detection and Diagnosis in engineering systems.* Marcel Dekker, New York/Basel/Hong Kong, 1998.
77. R. Ghabcheloo, A. Pascoal, C. Silvestre, and I. Kaminer. Nonlinear coordinated path following control of multiple wheeled robots with bidirectional communication constraints. *International Journal of Adaptive Control and Signal Processing*, 2006.

78. E. N. Gilbert. Capacity of burst-noise channels. *Bell Systems Technical Journal*, 39, 1960.
79. F. Giuletti, L. Pollini, and M. Innocenti. Autonomous formation flight. *IEEE Control Systems Magazine*, pages 34–44, 2000.
80. C. Guernez, J. P. Cassar, and M. Staroswiecki. Extension of parity space to nonlinear polynomial dynamic systems. *Proceeding of the 3th IFAC Symp. Fault Detection, Supervision and Safety of Technical Processes, SAFEPROCESS'97*, 2:861–866, 1997.
81. V. Gupta, R. M. Murray, and B. Hassibi. On the control of jump linear Markov systems with Markov state estimation. *Proceeding of American Control Conference*, pages 2893–2898, 2003.
82. H. Hammouri, P. Kabore, , and M. Kinnaert. A geometric approach to fault detection and isolation for bilinear systems. *IEEE Trans. Autom. Control*, 46(9):1451–1455, 2001.
83. S. Hashtrudizad, R. H. Kwong, and W. M. Wonham. Fault diagnosis in discrete-event systems: Framework and model reduction. *IEEE Transactions on Automatic Control*, 48(7):1199–1212, 2003.
84. S. Hashtrudizad, R. H. Kwong, and W. M. Wonham. Fault diagnosis in discrete-event systems: Incorporating timing information. *IEEE Transactions on Automatic Control*, 50(7):1010–1015, 2005.
85. S. Hashtrudizad and M. A. Massoumnia. Generic solvability of the failure detection and identification. *Automatica*, 35(5):887–893, 1999.
86. A. J. Healey. Application of formation control for multi-vehicle robotic minesweeping. *IEEE Conference on Decision and Control*, pages 1497–1502, 2001.
87. D. Hengy and P. M. Frank. Component failure detection via nonlinear state observers. *Proceeding of IFAC Workshop on fault detection and Safety in chemical plants*, 1:153–157, 1986.
88. R. Isermann. Process fault detection based on modeling and estimation methods: A survey. *Automatica*, 20(4):387–404, 1984.
89. R. Isermann. Experiences with process fault detection via parameter estimation. In S. G. Tzafestas, M. G. Singh, and G. Schmidts, editors, *System Fault Diagnostics, Reliability and Related Knowledge-based Approaches*, volume 1, pages 3–33. D. Reidel Press, Dordrecht, 1987.
90. A. Isidori, A. J. Kerner, C. Gori-Giorgi, and S. Monaco. Nonlinear decoupling via feedback: A differential geometric approach. *IEEE Transactions on Automatic Control*, 26:331–345, 1981.
91. B. Jiang, M. Staroswiecki, and V. Cocquempot. Fault diagnosis based on adaptive observer for a class of non-linear systems with unknown parameters. *International journal of control*, 77(4):415–426, 2004.
92. B. Jiang, M. Staroswiecki, and V. Cocquempot. Fault estimation in nonlinear uncertain systems using robust/sliding-mode observers. *IEE Proceedings on Control Theory Application*, pages 29–37, 2004.
93. Y. Jin, A. A. Minai, and M. M. Polycarpou. Cooperative real-time search and task allocation in UAV teams. *IEEE Conference on Decision and Control*, pages 7–12, 2003.
94. Z. Jin, V. Gupta, and R. M. Murray. State estimation over packet dropping networks using multiple description coding. *Automatica*, 42(9):1441–1452, 2006.
95. H. Jones. *Failure detection in linear systems.* PhD thesis, Massachusettes Inst. Technol, 1973.
96. J. Y. Keller, L. Summerer, M. Boutayeb, and M. Darouach. Generalized likelihood ratio approach for fault detection in linear dynamic stochastic systems with unknown inputs. *Int. J. Sys. Sci.*, 27(12):1231–1241, 1990.

97. Y. Kim and J. M. Watkins. A new approach for robust and reduced order fault detection filter design. *Proceeding of American Control Conference*, pages 1137–1142, 2007.
98. M. Kinnaert and Y. B. Peng. Residual generator for sensor and actuator fault detection and isolation- a frequency-domain approach. *International Journal of Control*, 61(6):1423–1435, 1995.
99. A. Knoll and J. Beck. Autonomous decision-making applied onto UAV formation flight. *Proceedings of the AIAA Conference on Guidance, Navigation, and Control*, 2006. AIAA Paper 2006-6202.
100. J. Korbicz, J. Koscielny, Z. Kowalczuk, and W. Cholewa. *Fault diagnosis, models, artifical intelligence, application*. Springer-Verlog, Berlin, 2004.
101. V. Krishnaswami and G. Rizzoni. Nonlinear parity equation residual generation for fault detection and isolation. *Proceeding of the* 2th *IFAC Symp. Fault Detection, Supervision and Safety of Technical Processes, SAFEPROCESS'94*, 1:317–322, 1994.
102. V. Krishnaswami and G. Rizzoni. Robust residual generation for nonlinear system fault detection and isolation. *Proc. IFAC Symp. Fault Detection, Supervision Safety Technical Processes*, pages 163–168, 1997.
103. T. R. Krogstad and J. T. Gravdahl. 6-DOF mutual synchronization of formation flying spacecraft. *Proceeding of IEEE Conference on Decision and Control*, 2006.
104. R. Krtolica, U. Ozguner, H. Chan, H. Gktas J. Winkelman, and M. Liubakka. Stability of linear feedback systems with random communication delays. *Int. J. Control*, 59(4):925–953, 1994.
105. C. R. Kube and H. Zhang. The use of perceptual cues in multi-robot box-pushing. *Proceeding of IEEE Conference on Robotics and Automation*, pages 2085–2090, 1996.
106. D. Kurabayashi, J. Ota, T. Arai, and E. Yoshida. Cooperative sweeping by multiple mobile robots. *Proceeding of IEEE Conference on Robotics and Automation*, pages 1744–1749, 1996.
107. P. Lawson. The terrestrial planet finder. *IEEE Proceedings of Aerospace Conference*, pages 2005–2011, 2001.
108. J. R. T. Lawton, R. W. Beard, and B. J. Young. A decentralized approach to formation maneuvers. *IEEE J. Robot. Autom.*, 19(6):933–941, 2003.
109. S. Li, D. Sauter, and C. Aubrun. Robust fault isolation filter design for networked control systems. *IEEE Conference on Emerging Technologies and Factory Automation*, pages 681–688, 2006.
110. Z. Q. Li, L. Ma, and K. Khorasani. A dynamic neural network-based reaction wheel fault diagnosis for satellites. In *Proc. IEEE International Joint Conference on Neural Networks (IJCNN-2006)*, Vancouver, Canada, July 2006.
111. J. Lofberg. Yalmip : A toolbox for modeling and optimization in MATLAB. In *Proceedings of the CACSD Conference*, Taipei, Taiwan, 2004.
112. X. Lou, G. C. Verghese, and A. S. Willsky. Optimally robust redundancy relations for failure detection in uncertain systems. *Automatica*, 22(3):333–344, 1986.
113. J. F. Magni and P. Mouyon. On residual generation by observer and parity space approaches. *IEEE Trans. Autom. Control*, 39(2):441–447, 1994.
114. V. Manikonda, P. O. Arambel, M. Gopinathan, R. K. Mehra, and F. Y. Hadaegh. A model predictive control-based approach for spacecraft formation keeping and attitude control. *Proceedings of the American Control Conference*, pages 4258–4262, 1999.
115. Z. Mao, B. Jiang, and P. Shi. H_∞ fault detection filter design for networked control systems modeled by discrete markovian jump systems. *IET Control Theory and Application*, 1(5):1336–1343, 2007.

116. T. Marcu. A multiobjective evolutionary approach to pattern recognition for robust diagnosis of process faults. *Proceeding of the 3th IFAC Symp. Fault Detection, Supervision and Safety of Technical Processes, SAFEPROCESS'97*, pages 1183–1188, 1997.
117. G. Marro. The geometric approach toolbox. *http://www3.deis.unibo.it/Staff/FullProf/GiovanniMarro/geometric.htm*, 2007.
118. M. A. Massoumnia. *A geoemtric approach to failure detection and identification in linear systems.* PhD thesis, Massachusettes Inst. Technol, 1986.
119. M. A. Massoumnia. A geometric approach to the synthesis of failure detection filters. *IEEE Trans. Autom. Control*, 31(9):839–846, 1986.
120. M. A. Massoumnia, G. C. Verghese, and A. S. Willsky. Failure detection and identification. *IEEE Transaction on Automatic Control*, 34(3):316–321, 1989.
121. T. W. Mclain and R. W. Beard. Trajectory planning for coordinated rendezvous of unmanned air vehicles. *Proceeding of the AIAA Conference on Guidance, Navigation, and Control*, AIAA 2000-4369, 2000.
122. D. McLean, S. Aslam-Mir, and H. Benkhedda. Fault detection and control reconfiguration in flight control. *IEE Colloquium on Fault Diagnosis and Control System Reconfiguration*, pages 1–1, 1993.
123. S. Mcllraith, G. Biswas, Dan Clancy, and Vineet Gupta. Hybrid systems diagnosis. *Proceedings of The Third International Workshop on Hybrid Systems: Computation and Control (HSCC 2000)*, pages 282–295, 2000.
124. R. K. Mehra and J. Peschon. An innovations approach to fault detection and diagnosis in dynamic systems. *Automatica*, 7:637–640, 1971.
125. M. Mesbahi and F. Y. Hadaegh. Formation flying control of multiple spacecraft via graphs, matrix inequalities and switching. *J. Guid., Contr., & Dyn.*, 24(2):369–377, 2001.
126. N. Meskin, T. Jiang, E. Sobhani, K. Khorasani, and C. A. Rabbath. A nonlinear geometric approach to fault detection and isolation in an aircraft nonlinear longitudinal mode. *Proceeding of American Control Conference*, pages 5771–5776, 2007.
127. N. Meskin and K. Khorasani. Fault detection and isolation of actuator faults in spacecraft formation flight. *Proceeding of the 45th IEEE Conference on Decision and Control*, pages 1159–1164, 2006.
128. N. Meskin and K. Khorasani. Fault detection and isolation in a redundant reaction wheels configuration of a satellite. *Proceeding of IEEE International Conference on Systems, Man, and Cybernetics*, pages 3153–3158, 2007.
129. N. Meskin and K. Khorasani. Fault detection and isolation of actuator faults in overactuated systems. *Proceeding of American Control Conference*, pages 2527 – 2532, 2007.
130. N. Meskin and K. Khorasani. Geometric approach to robust fault detection and isolation of Markovian jump systems. *Proceeding of American Control Conference*, pages 2822–2827, June 2008.
131. N. Meskin and K. Khorasani. Actuator fault detection and isolation for a network of unmanned vehicles. *IEEE Transactions on Automatic Control*, 54(4):835 840, 2009.
132. N. Meskin and K. Khorasani. Fault detection and isolation of discrete-time markovian jump systems with application to a network of multi-agent system having imperfect communication channels. *Automatica*, 45(9):2032–2040, 2009.
133. N. Meskin and K. Khorasani. Unobservability subspaces for continuous-time Markovian jump systems with application to fault diagnosis. *Proceeding of the 7th IFAC Symposium on Fault Detection, Supervision and Safety of Technical Processes*, pages 953–958, 2009.

134. N. Meskin and K. Khorasani. A geometric approach to fault detection and isolation of continuous-time Markovian jump systems. *IEEE Transactions on Automatic Control*, 55(6):1343–1357, 2010.
135. N. Meskin, K. Khorasani, and C. A. Rabbath. A hybrid fault detection and isolation strategy for nonlinear systems in presence of large environmental disturbances. *IET Control Theory and Applications.* to appear.
136. N. Meskin, K. Khorasani, and C. A. Rabbath. Fault detection and isolation strategy for a network of unmanned vehicles in presence of large environmental disturbances,. *AIAA Guidance, Navigation, and Control Conference, Chicago*, 2009. AIAA-2009-5656.
137. N. Meskin, K. Khorasani, and C. A. Rabbath. Fault diagnosis in a network of unmanned aerial vehicles with imperfect communication channels. *AIAA Guidance, Navigation, and Control Conference, Chicago*, 2009. AIAA-2009-6100.
138. N. Meskin, K. Khorasani, and C. A. Rabbath. A hybrid fault detection and isolation strategy for a network of unmanned vehicles in presence of large environmental disturbances. *IEEE Transactions on Control System Technology*, 18(6):1422–1429, 2010.
139. S. Narasimhan and G. Biswas. Model-based diagnosis of hybrid systems. *IEEE Transactions on Systems, Man and Cybernetics-Part A: Systems and Humans*, 37(3):348–361, 2007.
140. K. S. Narendra and S. S. Tripathi. Identification and optimization of aircraft dynamics. *J. Aircraft*, 10(4):193–199, 1973.
141. P. Ogren, M. Egerstedt, and X. Hu. A control Lyapunov function approach to multi-agent coordination. *IEEE Trans. on Robotics and Automation*, 18:847–851, 2002.
142. J. H. Park and G. Rizzoni. An eigenstructure assignment algorithms for the design of fault detection filters. *IEEE Trans. Autom. Control*, 39(7):1521–1524, 1994.
143. J. H. Park and G. Rizzoni. A new interpretation of the fault-detection filter, 1: Closed-form algorithm. *International Journal of Control*, 60(5):767–787, 1994.
144. J. H. Park, G. Rizzoni, and W. B. Ribbens. On the representation of sensor faults in fault-detection filters. *Automatica*, 30(11):1793–1795, 1994.
145. L. E. Parker. Alliance: An architecture for fault tolerant multirobot cooperation. *IEEE Transaction on Robotics and Automation*, 14(2):220–240, 1998.
146. K. Patan and T. Parisini. Identification of neural dynamic models for fault detection and isolation: the case of real sugar evaporation process. *Journal of Process Control*, 15(1):67–79, 2005.
147. R. J. Patton. Robust fault detection using eigenstructure assignment. *Proc. 12th IMACs World Congress on Scientific Computation*, pages 431–434, 1988.
148. R. J. Patton and J. Chen. Observer-based fault detection and isolation: robustness and applications. *Contr. Eng. Practice*, 5(5):671–682, 1997.
149. R. J. Patton, P. M. Frank, and R. N. Clark, editors. *Fault diagnosis in dynamic systems, theory and applications.* Control Engineering series, Prentice-Hall, New York, 1989.
150. R. J. Patton, P. M. Frank, and R.N. Clark. *Issues of fault diagnosis for dynamic systems.* Springer-Verlog, London, 2000.
151. R. J. Patton and Chen J. A review of parity space approaches to fault diagnosis. *Proceeding of the first IFAC Symp. Fault Detection, Supervision and Safety of Technical Processes, SAFEPROCESS'91*, 1:239–255, 1991.
152. C. D. Persis and A. Isidori. Fault detection and isolation for state affine systems. *European Journal of Control*, 6:290–294, 2000.
153. C. D. Persis and A. Isidori. On the observability codistributions of a nonlinear system. *Systems and Control Letters*, 40:297–304, 2000.

154. C. D. Persis and A. Isidori. A geometric approach to nonlinear fault detection and isolation. *IEEE Transactions on Automatic Control*, 46(6):853–865, 2001.
155. C. D. Persis and A. Isidori. On the design of fault detection filters with game-theoretic-optimal sensitivity. *International Journal of Robust and Nonlinear Control*, 12:729–747, 2002.
156. L. Pollid, F. Giulietti, and M. Innocenti. Robustness to communication failures within formation flight. *IEEE Proceedings of the American Control Conference*, pages 2860–2866, 2002.
157. Wei Ren and Randal W. Beard. A decentralized scheme for spacecraft formation flying via the virtual structure approach. *AIAA Journal of Guidance, Control and Dynamics*, 27(1):73–82, 2004.
158. M. Sampath, R. Sengupta, S. Lafortune, and K. Sinnamohideen. Failure diagnosis using discrete-event models. *IEEE Transactions on Control System Technology*, 4(2):105–124, 1996.
159. D. P. Scharf, F. Y. Hadaegh, and S. R. Ploen. A survey of spacecraft formation flying guidance and control (part I): Guidance. *Proceedings of the American Control Conference*, pages 1733–1739, 2003.
160. D. P. Scharf, F. Y. Hadaegh, and S. R. Ploen. A survey of spacecraft formation flying guidance and control (part ii): Control. *Proceedings of the American Control Conference*, pages 2976–2985, 2004.
161. H. Schaub and J. L. Junkins. *Analytical Mechanics of Space Systems*. AIAA Educational Series, 2003.
162. P. Seiler and R. Sengupta. A bounded real lemma for jump systems. *IEEE Trans. Autom. Control*, 48(9):1651–1654, 2003.
163. P. Seiler and R. Sengupta. An H_∞ approach to networked control. *IEEE Trans. Autom. Control*, 50(3):356–364, 2005.
164. R. Seliger and P. M. Frank. Fault diagnosis by disturbance decoupled nonlinear observer. *Proceedings of IEEE Conference on Decision and Control*, pages 2248–2253, 1991.
165. R. Seliger and P. M. Frank. Fault diagnosis by disturbance decoupled nonlinear observers. *Proc. of the* 30th *IEEE Conf. on Decision and Control*, pages 2248–2253, 1991.
166. R. Seliger and P. M. Frank. Robust observer-based fault diagnosis in nonlinear uncertain systems. In R. J. Patton, P. M. Frank, and R. N. Clark, editors, *Issues of fault diagnosis for dynamic systems*. Springer Verlag, Englewood Cliffs, NJ, 2000.
167. S. Shankar, S. Darbha, and A. Datta. Design of a decentralized detection filter for a large collection of interacting LTI systems. *Mathematical Problems in Engineering*, 8(3):233–248, 2002.
168. P. Shi, E. K. Boukas, and R. K. Agarwal. Control of Markovian jump discrete-time systems with norm bounded uncertainty and unknown delay. *IEEE Trans. Autom. Control*, 44(11):2139–2144, 1999.
169. S. E. Shladover and et al. Automatic vehicle control developments in the path program. *IEEE Transactions on Vehicular Technology*, 40(1):114–130, 1991.
170. A. Shumsky. Robust residual generation for diagnosis of nonlinear systems: parity relation approach. *Proceeding of the* 3th *IFAC Symp. Fault Detection, Supervision and Safety of Technical Processes, SAFEPROCESS'97*, 2:867–872, 1997.
171. S. Simani, C. Fantuzzi, and R. J. Patton. *Model-based fault diagnosis in dynamic systems using identification techniques*. Springer, 2003.
172. R. S. Smith and F. Y. Hadaegh. Parallel estimation and communication in spacecraft formations. *IFAC World Congress*, 2005.
173. E. Sobhani, K. Khorasani, and S. Tafazoli. Dynamic neural networkbased estimator for fault diagnosis in reaction wheel actuator of satellite attitude control system. *Proc. IEEE International Joint Conference on Neural Networks (IJCNN-2005)*, pages 2347–2352, 2005.

174. J. L. Speyer. Computation and transmission requirements for a decentralized linear-quadratic-gaussain control problem. *IEEE Transaction on Automatic Control*, 24(2):226–269, 1979.
175. R. Sreedhar, B. Fernandez, and G. Y. Masada. Robust fault detection in nonlinear systems using sliding mode observers. *Proceedings of IEEE Conference on Control Applications*, pages 716–721, 1993.
176. R. Sun, F. Tsung, and L. Qu. Combining bootstrap and genetic programming for feature discovery in diesel engine diagnosis. *International Journal of Industrial engineering*, 11(3):273–281, 2004.
177. H.A. Talebi and R.V. Patel. A neural network-based fault detection scheme for satellite attitude control systems. *Proc. IEEE Control Applications Conf.*, pages 1293–1298, 2005.
178. H.A. Talebi and R.V. Patel. An intelligent fault detection and recovery scheme for reaction wheel actuator of satellite attitude control systems. *Proc. IEEE Control Applications Conf.*, pages 3282–3287, 2006.
179. Abhishek Tiwari, Jimmy Fung, John M. Carson, Raktim Bhattacharya, and Richard M. Murray. A framework for Lyapunov certificates for multi-vehicle rendezvous problems. *IEEE Proceedings of the American Control Conference*, pages 5582–5587, 2004.
180. N. Tudoroiu and K. Khorasani. Satellite fault diagnosis using a bank of interactive Kalman filters. *IEEE Transactions on Aerospace and Electronic Systems*, 43:1334–1350, 4 2007.
181. S. G. Tzafestas and K. Watanabe. Modern approaches to system/sensor fault detection and diagnosis. *Journal A. IRCU Lab*, 31(4):42–57, 1990.
182. T. Vidal, M. Ghallab, and R. Alami. Incremental mission allocation to a large team of robots. *Proceeding of IEEE Conference on Robotics and Automation*, pages 1620–1625, 1996.
183. N. Viswanadham and K. D. Minto. Robust observer design with application to fault detection. *Amer. contr. Conf.*, pages 1393–1399, 1988.
184. H. Wang, C. Wang, H. Gao, and L. Wu. An LMI approach to fault detection and isolation filter design for Markovian jump system with mode-dependent time-delays. *Proceedings of the American Control Conference*, pages 5686–5691, 2006.
185. P. K. C. Wang and F. Y. Hadaegh. Coordination and control of multiple microspacecraft moving in formation. *Journal of the Astronautical Sciences*, 44(3):315–355, 1996.
186. K. Watanabe and D. M. Himmelblau. Instrument fault detection in systems with uncertainties. *Int. J. Sys. Sci.*, 13(2):137–158, 1982.
187. M. Weerasinghe, J. B. Gomm, and D. Williams. Neural networks for fault diagnosis of a nuclear fuel processing plant at different operating points. *Control Engineering Practice*, 6(2):281–289, 1998.
188. J. E. White and J. L. Speyer. Detection filter design: spectral theory and algorithms. *IEEE Transaction on Automatic Control*, 32(7):593–603, 1987.
189. A. S. Willsky and H. L. Jones. A generalized likelihood approach to the detection and estimation of jumps in linear systems. *IEEE Trans. Autom. Control*, 21:108–121, 1976.
190. M. Witczak. *Modeling and estimation strategies for fault diagnosis of nonlinear system: from analytical to soft computing approach.* Springer-Verlog, 2007.
191. J. D. Wolfe, D. F. Chichka, and J. L. Speyer. Decentralized controllers for unmanned aerial vehicle formation flight. *Proceedings of the AIAA Conference on Guidance, Navigation, and Control*, 1996. AIAA Paper 96-3833.
192. W. M. Wonham. *Linear multivariable control: A geometric approach.* Springer-Verlag, New York, third edition, 1985.
193. N. E. Wu and Y. Y. Wang. Robust failure detection with parity check in filtered measurements. *IEEE Trans. Aerosp. Electron. Syst.*, 31(1):489–491, 1995.

194. Q. Wu and M. Saif. Neural adaptive observer based fault detection and identification for satellite attitude control systems. *American Control Conference*, pages 1054–1059, 2005.
195. L. Xiao, A. Hassibi, and J. How. Control with random communication delays via a discrete-time jump system approach. *Proceeding of American Control Conference*, pages 2199–2204, 2000.
196. J. Xing and J. Lam. Fixed-order robust H_∞ filter design for Markovian jump systemswith uncertain switching probabilities. *IEEE Trans. Signal Process.*, 54(4):1421–1430, 2006.
197. J. Xiong and J. Lam. Stabilization of linear systems over networks with bounded packet loss. *Automatica*, 43:80–87, 2007.
198. J. Xiong, J. Lam., H. Gao, and D. W. C. Ho. On robust stabilization of Markovian jump systems with uncertain switching probabilities. *Automatica*, 41(5):897–903, 2005.
199. H. Yamaguchi. Adaptive formation control for distributed autonomous mobile robot groups. *Proceeding of IEEE Conference on Robotics and Automation*, pages 2300–2305, 1997.
200. X. Yan and C. Edwards. Nonlinear robust fault reconstruction and estimation using a sliding mode observer. *Automatica*, 43(9):1605–1614, 2007.
201. H. Yang and M. Saif. Nonlinear adaptive observer design for fault detection. *Proceedings of the American Control Conference*, pages 1136–1139, 1995.
202. Y. Yang, A. A. Minai, and M. M. Polycarpou. Decentralized cooperative search by networked UAVs in an uncertain environment. *IEEE Proceedings of the American Control Conference*, pages 5558–5563, 2004.
203. H. Ye and S. X. Ding. Fault detection of networked control systems with networked-induced delay. *Proceeding of the 8th International Conference on Control, Automation, Robotics and Vision*, pages 294–297, 2004.
204. J. Yuh. Underwater robotics. *Proceedings of the IEEE Conference on Robotics and Automation*, pages 932–937, 2000.
205. P. Zhang, S. X Ding, P. M. Frank, and M. Sader. Fault detection of networked control systems with missing measurements. *Proceeding of the 5th Asian Control Conference*, pages 1258–1263, 2004.
206. Y. Zheng, H. Fang, and Y. Wang. Kalman filter based FDI of networked control system. *Proceeding of the 5th World Congress on Intelligent Control and Automation*, pages 1330–1333, 2004.
207. M. Zhong, S. X. Ding, J. Lam, and H. Wang. An LMI approach to robust fault detection filter for uncertain LTI systems. *Automatica*, 39:543–550, 2003.
208. M. Zhong, S. X. Ding, and B. Tang. An LMI approach to robust fault detection filter for discrete-time systems with model uncertainty. *Proceeding of the 40th conference on decision control*, pages 3613–3618, 2001.
209. M. Zhong, H. Ye, P. Shi, and G. Wang. Fault detection for Markovian jump systems. *IEE Proc. Control Theory Application*, 152(4):397–402, 2005.

Index